Migration, Diaspora and Information Technology in Global Societies

Routledge Research in Information Technology and Society

Migration, Diaspora and Information Technology in Global Societies

Edited by Leopoldina Fortunati, Raul Pertierra and Jane Vincent

Routledge
Taylor & Francis Group
NEW YORK   LONDON

First published 2012
by Routledge
711 Third Avenue, New York, NY 10017

Simultaneously published in the UK
by Routledge
2 Park Square, Milton Park, Abingdon, Oxon OX14 4RN

*Routledge is an imprint of the Taylor & Francis Group,
an informa business*

Typeset in Sabon by IBT Global.
Printed and bound in Great Britain by
CPI Antony Rowe, Chippenham, Wiltshire

Library of Congress Cataloging-in-Publication Data
Migration, diaspora, and information technology in global societies / edited
    by Leopoldina Fortunati, Raul Pertierra, and Jane Vincent. — 1st ed.
        p. cm. — (Routledge research in information technology and society ; 12)
    Includes bibliographical references and index.
    1. Information technology—Social aspects.   2. Internet and immigrants.
3. Identity (Psychology) and mass media.   4. Communication and
technology.   I. Fortunati, Leopoldina.   II. Pertierra, Raul, 1941–
III. Vincent, Jane, 1956–
    HM851.M54 2012
    303.48'33086912—dc23
    2011025196

ISBN: 978-0-415-88709-0 (hbk)
ISBN: 978-0-203-14860-0 (ebk)

Contents

THEME 3
Looking at the Migrations and Diasporas Through the
Lens of the New Media

THEME 4
Religion, Mobility and Social Policies: How Migrants' Use
of the New Media Is Shaping Society

THEME 5
A Case Study: China, Its Internal Migrations, Diasporas and Expatriates

Figures and Tables

FIGURES

TABLES

Foreword

*Leopoldina Fortunati, Raul Pertierra
and Jane Vincent*

The world today is being consistently transformed by internal and external migration and, indeed, if the external migrants were to constitute a country it would most likely rank among the first ten globally. This book aims to do justice to the great diversities of the multicultural, multilingual and multiethnic population of migrants. Coming from elsewhere, and using ICTs, migrants may choose to escape the inertia and fixity of their new local culture or they may interweave it with their own cultural differences into one new fabric. The everyday life experiences explored in this volume derive from understanding and comparing the activities of migrants in these different cultures in which they have built their new lives.

As the authors of this book describe, ICTs are the new ecosystem in a migrant's life, which span the elsewhere and the here, the present and the absent. Social and familial bonds are maintained, developed and denied; emotion and affections are experienced, memories are more or less recorded, the old and the new cultural identities negotiated, the masculinity and femininity continuously elaborated, the generations shaped and reshaped and different languages practiced: in sum, ICTs are the environments in which migrants, as the new locals, live their everyday life.

This book retraces migrants' practices of using these technologies on all the continents—Europe, Asia, Australia, the Americas and Africa—but it also deals with relevant theoretical issues which constitute problems still open in migration studies. Looking at migrants through the lens of ICT use offers new knowledge about a body of humankind on the move with a mobile phone in their pocket or access to the Internet to guide them, keep them in touch and maintain their life wherever they may be.

Preface
Mediating a Restless World

Daniel Miller

In a way it is strange that something so enormous and profound can at first sound so bland. To announce that this is "a book on media and migration" as a string of words does not of itself jolt us out of our attention to the everyday concerns we are busy with. But it should, and it does once we engage with our own knowledge and imagination of what these words imply. Let us start with media. We begin with a relatively obvious image: journalism and newspapers. Already we can start to appreciate that the role media plays in relation to migration is not merely reflective but determinant. In the U.K., where I live, the topic of migration, certainly immigration, is seen as central to politics, a key issue in the election of governments. It is perhaps *the* topic that generates right wing political activism and one which certainly frightens the rest of the political spectrum. Increasingly it seems to be the single issue that determines how we are governed, certainly in those cases where a party focused upon immigration (although relatively small) holds the balance of power, as is the case in several countries. But let us be clear. It is not immigration itself that is responsible for this state of affairs, it is the *media's* portrayal of immigration. It is the exploitation of this issue by the media. If there was no media interest then generally migration would certainly be significant but not so central to the determination of political outcomes. The consequences of this are by no means limited to politics. Being a migrant, the way people look at you, respond to you and whether or not they accept you seems to be beholden to this vastly powerful force—the media. So the first topic of this volume, the way the media represents migrants and the consequences of this representation, is hugely important not just to migrants but to the whole governance of the world.

But that is only the tip of an iceberg; this initial discussion then raises the question of what we mean when we use the term media. What has been implied so far is actually a rather old fashioned concept of what we might now call "old media"—the way the newspapers and television report to a passive audience news about migration. But most of the papers in this book are trying to encompass the state of the art in media, the new and very new media. Today we live in an unprecedented situation, described by Madianou and myself (2011) as Polymedia. Within the last two years very

ordinary people around the world, living not just in cities but in villages
and small towns too, find that their relationship to media is entirely transformed once they have the money for computer access and phone services.
Firstly they have a plethora of different media available to them from social
networking sites, with YouTube, with blogging, Web cam, text messaging
and with those infrastructures we so easily and so quickly take for granted,
such as search engines. Secondly the issue of price moves from foreground
to background because once the cost of the computer is dealt with there
may be no actual cost to any message. This means that very ordinary low
income people (the kind of people who migrate as workers) are developing
an entirely new relationship to the media. Not as passive objects reported
on by the media, but as the creators of media themselves, through blogs,
YouTube and so on, and in their own communication to their Facebook
friends. This is the second subject of this book, which complements the
first: migrants as the makers of news, not just the objects of news.

Yet both of these considerations which seem hugely important at one
level almost pale into insignificance when we start to pay attention to
a third implication of this relationship between migration and media.
What aspect of the media matters most to the migrants themselves?
What happens when we stop thinking of media in terms of any kind of
news and turn instead to the social relationships of ordinary people?
When Mirca Madianou and myself decided to start research on a project
investigating long distance communication between migrants and their
left-behind families, we applied for grants under the following argument:
There are of course many ways you can study migrants and many valuable topics that help us understand this situation. But what if instead of
deciding for yourself what matters in migration you turn to the migrants
themselves? What would they say was the single most important thing
in the determination of their welfare, as to whether they would feel miserable or relatively comfortable with their situation? We suggested that
what matters most to these migrants is exactly the same as what matters
most to non-migrants. That is, their relation to a few core people whom
they love: their children, parents, lovers, siblings and best friends. If we
read a novel or see a film about migration that touches us it is very likely
because these relationships of love will be central to the portrayal of
those lives.

Once again it is when we start to use our imagination to think of what
is at stake that media moves from background to foreground. The reason
we are working on the way Filipina mothers use media to contact their left-
behind children is because we realized that today the very degree to which
a mother can consider herself to actually be a mother in practice depends
almost entirely on those media. They have moved from a situation of being
barely in contact at all to being able to check on every detail of homework
and their daughter's boyfriends through being in touch many times a day.
So it seemed reasonable to spend at least some time on that which makes

the most difference—that which matters most to migrants themselves—
and once again this is now dependent upon media.

These are not three separate categories of academic enquiry. In the contemporary world of Polymedia it is not just that we can now blog, Facebook, text, Web cam or simply phone, for these things have not replaced
newspapers and television but live alongside them. So as is clear in several
of the chapters of this book we also need to pay attention to the points
of intersection between these macro forces of news and the intimate lives
mediated via new media communication, which come into play once we
start to consider issues such as education and knowledge and privacy.

All of this is what explodes in front of us when we examine one half
of the equation that links media to migration. When we turn to the other
half, that of migration itself, the world seems even larger and even more
restless. I recall my first ethnographic work in Trinidad over twenty years
ago. I was astonished then that the majority of families I was studying, and
therefore by implication the majority of families in the country as a whole,
included a family member living abroad. Not a distant family member but
a nuclear family member—that is, a parent, child or sibling. The normal,
indeed the typical family was a migrant family. Gradually I came to realize
that this was not something special to Trinidad but true in country after
country, some of which now have their whole national economy more or
less dependent upon these external relationships. It means that we have to
stop thinking of migration as a special case, an aberration from the norm,
and once again move the background to the foreground, so that our imagination of the ordinary and the typical starts from the situation of migration. This is not a book about a special case; it is a book about ordinary
life. Still, we are only partway towards the reorientation of our vision that
is required, not least because the headline migration that we tend to think
about is that of international movement. However, migration has always
been and continues to be also a central aspect of that which happens inside
countries. When it comes to the centrality of media it really does not matter
whether the family member is in another country or another part of one's
own country, what matters is that one's relationship has now become a
media and mediated relationship, not a face-to-face relationship. This point
emerges with particular clarity in the series of papers on China that form
the final section of this book.

Finally there is a need to confront populations of this world as they
actually are. So much of the literature on migration and so much of the
literature on media tends to focus upon those concerns that dominate the
wealthiest, most metropolitan regions, in which academia is differentially
situated, especially Western Europe and North America. It is their media
and their immigrants that dominate so far. But most people in the world
live in China and South Asia. It is the latest media statistics that really bring
this home to us. When we realize that there are four billion mobile phones
in the world but that three billion of these are found in what have been

called the developing countries, we realize where most of the world actually is. It is not just movements from backgrounds to foregrounds then that we are in need of but also a different vision laterally across the world.

All of this represents a considerable challenge to the very idea of a book on media and migration. How would we even start to envisage a volume that can encompass this huge restless planet, the proliferation of media and the actual nature and consequence of migration in all its vastness? Clearly we would need an edited collection rather than a simple monograph because to address the whole range from infrastructure to the intimate means different styles and different points of focus. Similarly there are differences as we try to express the range, from how migration is represented in media to how people use the media, to include internal migration within countries as well as transnational diasporas. Furthermore in a situation where we cannot hope to include all migration we need a judicious choice of example, preferably a selection in which the major populations such as those of China become the key case studies and not simply an addition to the rest.

The implications of this preface are that the book which follows is not simply a collection but in some ways a reflection of the only kind of academic response that actually contends with the true implications of that simple phrase—media and migration. It does indeed use China as its primary case study; it does include papers that range from the representation of migrants in its first theme to the ways which migrants do, and interestingly sometimes do not, actually use media to maintain their contacts to their primary relationships and the people they love. It includes papers that make us appreciate the centrality of states and markets and how we need to bridge between these macro forces and the personal and human worlds of particular migrants. It includes papers on the subtleties of the media itself, including language. It represents a wider range of academic methods in investigating these phenomena, from the patient immersion in Jamaican life represented by high quality ethnography to the comparative study of European data sets. Finally it recognizes that these themes touch on almost any wider social parameter with which social sciences need to engage: from gender to generation to ethnicity to kinship, from the economic to the political to the religious. There is a limit to what any one volume can hope to accomplish, but at the very least this particular volume has the ambition to demonstrate what it is we need to encompass.

REFERENCES

Madianou, Mirca, and Daniel Miller. 2011. *Technologies of love: Migration and the Polymedia revolution*. London: Routledge.

Acknowledgments

The editors wish to thank Daniel Miller for writing the preface and the authors for contributing chapters written especially for this new volume. This book is closely related both to the scientific initiative and the life of COST 298—the European network named "Participation in the Broadband Society"—and to the Pordenone Group of Scholars who emerged from this network as a new source of reference for several scientific communities and initiatives in the field of ICT studies.

The members of the COST Action 298 and their Chair Bartolomeo Sapio and Vice Chair Tomaz Turk, together with the University of Udine, its Doctoral Programme in Multimedia Communication and the Faculty of Education are thanked for sponsoring and encouraging participation in the international seminar at which this book project was first mooted.

COST—the acronym for European Cooperation in Science and Technology—is the oldest and widest European intergovernmental network for cooperation in research. Established by the Ministerial Conference in November 1971, COST is presently used by the scientific communities of 36 European countries to cooperate in common research projects supported by national funds. The funds provided by COST—less than 1% of the total value of the projects—support the COST cooperation networks (COST Actions) through which, with EUR 30 million per year, more than thirty thousand European scientists are involved in research having a total value which exceeds EUR 2 billion per year. This is the financial worth of the European added value that COST achieves. A "bottom up approach"

(the initiative of launching a COST Action comes from the European scientists themselves), "à la carte participation" (only countries interested in the Action participate), "equality of access" (participation is open also to the scientific communities of countries not belonging to the European Union) and "flexible structure" (easy implementation and light management of the research initiatives) are the main characteristics of COST. As a precursor of advanced multidisciplinary research, COST has a very important role for the realization of the European Research Area (ERA) anticipating and complementing the activities of the Framework Programmes, constituting a "bridge" towards the scientific communities of emerging countries, increasing the mobility of researchers across Europe and fostering the establishment of "Networks of Excellence" in many key scientific domains such as Biomedicine and Molecular Biosciences; Food and Agriculture; Forests, their Products and Services; Materials, Physical and Nanosciences; Chemistry and Molecular Sciences and Technologies; Earth System Science and Environmental Management; Information and Communication Technologies; Transport and Urban Development; Individuals, Societies, Cultures and Health. It covers basic and more applied research and also addresses issues of pre-normative nature or of societal importance.

ESF provides the COST office through an EC contract.
Web: http://www.cost.eu

COST is supported by the EU RTD Framework Programme.

COST 298 – Participation in the Broadband Society.

Introduction
Migrations and Diasporas— Making Their World Elsewhere

Leopoldina Fortunati, Raul Pertierra and Jane Vincent

The appropriation of the new media by migrants has changed the way in which today people migrate, move, negotiate their personal and national identity and make strategies to deal with new cultures. Central among the main causes of the diffusion, adoption and domestication of ICTs by migrants are the globalization and the development of the broadband society. Their mutual ontology is proving to become one of the leading forces for political, cultural and economic transformation in the contemporary world, and in this chapter the topics that expand upon and explore this debate are set out in a series of themed discussions concluding with an overview of the contents of the five sections of the book.

In this book the various chapters investigate how, and with which social effects, information and communication technologies are used by specific social groups who are migrants and for this reason are considered to be strangers in a host society. We use the term migrants in a general meaning, although the politics of migrations distinguishes between emigrants who are generally pushed by war or starvation, immigrants who are instead pulled by the will to find a job or by freedom, and émigrés who are those that were obliged to emigrate because their land was occupied (McLean 1996, 321). This volume looks in particular at how the practices and the trajectories of use of the technology influence the important phenomenon of migrations and diasporas, which could be seen as the emblem of the late modernity. The multifaceted role that these technologies perform in the dislocation of consistent groups of people is explored here. We question how the technologies allow migrants to remain in contact with their communities of origin, to construct or enhance the relational sphere with local communities, to discuss their national identity, the identity of native communities and of the global diaspora they encounter. We further explore how these technologies might be used to meet in the "virtual common world" migrants of other ethnic groups and/or nationalities with whom migrants share the identity of "strangers" and from whom it is often possible to have information about job possibilities and more. Paradoxically, these new technologies of communication exacerbate but also attenuate the experiences of solitude and social exclusion or deprivation both among migrants and those who remain behind.

FROM INTEGRATION TO CO-CONSTRUCTION

The book analyzes the different ways in which migrants, together with natives, contribute to co-construct contemporary societies and cultures by drawing on diverse traditions. In turn, it explores the different ways in which contemporary societies contribute to co-construct migrations and to being native in their own country of residence. These dynamics between natives and migrants and their mutual coming together are still poorly understood, partly because the social sciences have failed to develop adequate theoretical models for understanding migration and producing communitarian policies capable of overcoming different national paradigms (Heckmann and Schnapper 2003). For this reason, we prefer to under-use the term "integration" because of its heavy and often assumed description of the cultural adaptation process of migrants in a taken-for-granted community of practices of the host society. The host society, as well as migrants, is considered as a kind of fixed cultural reality which is not subject to any internal or external process of change and with which "the other" has to come to terms. Although widespread, the term integration does not reach a large consensus: sometimes it is connected to desegregation, other times to the attempt of bringing minorities' cultures into the mainstream of cultures and their social structures, including rights and services.

Our stance in this book is that one should talk of co-construction, which starts from the presupposition that each society is a dynamic system which meets and maybe clashes with other cultures, but in so doing enriches itself and consequently changes. The co-construction is a process in which locals and migrants give life to a different society in which both cultures are considered in their interaction and where both cultures have the concrete possibility to learn, reflect on and modify particular aspects of their everyday life. This concept develops the term "cultural co-traditions" advanced by Ferrarotti (1999, 158), in which the acceptance and coexistence of different cultures in a society is seen as the only way out of the problems posed by migrations. In particular, our glimpse is on the socio-technical systems that migrants and natives co-construct inside contemporary societies.

According to Scott and Marshall (2005, 155–156), traditional social studies on migrations and diasporas considered issues such as unidirectional flows, uprooting of migrants from their societies and cultures of origin and assimilation via the melting pot into the new host culture. These two scholars assert that contrary to this view, post-modern social studies usually focus on the improvements in transport and communications which have made it possible for migrants to implement their own distinctive identities, lifestyles and economic ties. This second wave of studies, of course, also reconsiders the question of modern nation-states by replacing the rigid territorial nationalism with the notion of shifting and contested boundaries, creating new terms such as "imagined communities" (Anderson 1983) and "global ethnospaces" (Appadurai 1996) to describe transnational influences. It is evident that there is a tension and even a counter-position between these two approaches: traditional sociologists of migrations—continue Scott and Marshall—criticize post-modern scholars for the creation of abstruse theoretical terminology, their "apparent disregard for numbers and generalizations, and a tendency to ignore earlier sociological studies of migration (especially where these document complex structures of opportunity and migrants networks in ways which prefigure the new diaspora studies themselves)" (Scott and Marshall 2005, 155). Scott and Marshall are instead convinced that the new wave of studies on migration and diaspora is able to detail "the complexity, diversity, and fluidity of migrant identities and experiences in a more realistic way than did the older mechanistic theories and models of international migration" (2005, 155).

In our case, the expression "migrants in late modernity" aims to underline the perspective that sees migrants playing a more or less powerful role in the co-construction of contemporary societies. However, our approach is not focused on structural economic and political influences upon or on the part of migrants. Instead, we aim to contribute to the understanding of some of the relevant social problems that current societies are faced with, such as identity, social cohesion and structure of social stratification, by examining these topics from the perspective of ICT use by migrants. The migrants in fact do so by experiencing and developing the information and communication technologies, that is, by domesticating them in a way that is peculiar to their needs, both the means of information and communication they use and the communicative act (Miller and Horst 2006; Donner 2008). This emphasis is both methodological as well as substantive. The contributions of migrants as well as the growing significance of media are recognized. Thus in this book we focus on the communicative, linguistic, emotional and technological dimensions experienced by migrants in their relationship with the new (and the old) media. However, this approach is not completely new but has an old lineage.

A century ago, Sombart argued (1902/1927) that migrants historically gave a great impulse to technological innovations. Some leading examples are Einstein, Fermi, Marconi, Von Neumann, Wittgenstein and Gödel, who were all migrants. Far from hostile circumstances in their country of origin, they developed their creativity and imagination in a new host country. Migrants' contribution applies as much to the Arts as to the Sciences. African musicians, Jewish refugees, Soviet dissidents and Asian intellectuals have not limited themselves to "integrating" in the host culture and society but they have enriched and reoriented many contemporary societies and cultures. An example of the reorientation of the cultural identity of a country is that provided by Gerard Goggin (2008): Australia. This country, because of the growing migration from Japan and China, has reoriented its identity towards the Asia Pacific instead of towards the old Europe and in particular towards the U.K.

GLOBAL TRENDS OF MIGRATIONS

There are four elements which in a certain sense characterize contemporary migrations:

1. Their dimensions
2. A new gender balance among migrants
3. The fact that today migrants can count on the availability of ICTs as powerful tools to mediate their trajectory of life
4. The parallel development of other figures of contemporary mobility such as tourists, travelers, commuters and so on

In regard to the first element, the world has never registered so high a number of migrants. It is as if humankind decided to put itself on the move. As Simmel (1998, 568) had already noticed, the extraordinary increase of the need for the differences in the modern individual brings both the differentiation of the personal and social existences and the need for mobility. According to William Lacy Swing (responsible for the International Organization for Migration),[1] almost a billion of the people in the world are migrants: 214 million have migrated abroad and 700 million inside their country.[2] The account of migrations today is changing the profile of humankind and popularizes an experience which in the past has involved only some sections of worldwide population. As Daniel Miller writes in the Preface (this volume), "This is not a book about a special case, it is a book about ordinary life." Of course the motivations that are behind these impressive migrations are numerous: wars, starvation, political persecutions, lack of a job, poverty and so on.

As to the second element, differently from the past women today account for half of the global migrants. This important change in migrants' gender resonates with the increase of feminine power at an international level and has of course transformed both women's conditions of life and the social morphology of the territories from which these women migrated (see Heather Horst's chapter in this volume and Dustmann 2005a, 2005b, 2008). Traditionally women have always been more sedentary than men because of their family roles, although a certain number of women have experienced various types of travel and a high number at least a honeymoon (Corsi 1999). There have always been those who remained at home or who left to follow after the men of the family after they were well established in the new country and could call their wives and children to join them. Now, in many cases, it is the women who constitute the first wave of migrants (Pingol 2001) because the labor market specifically requires their labor force, as in the case of caregivers, or they migrate together with the men: this is in tune also with the increasing numbers of women who travel or do tourism alone (Silvestre and Valerio 1999). Among other things the demands of housework in families of industrialized countries whose women are in the workforce, coupled with the reduction of social services, have created the unprecedented need for a feminine migration (Hochschild and Ehrenreich 2003). Newly emerging economies such as Hong Kong, Singapore and the Middle East have added significantly to this need for female migration. What are the consequences for families in both receiving and offering feminine migrant labor? What specific empowering processes for women occur, what redesignation of their gender identity and narratives and what new communicational practices does it generate? Does this strong presence of women among migrants have specific effects on the use of ICTs? If so, what are they? The Philippines, for example, has one of the highest rates of worker-migrants (Pertierra 2006), the majority being women, and it is worthy of note that recently an online counseling service has become available for the families of migrants in order to alleviate the problems of distant parenting. This change is contradictory because it has created many problems in the life of these women and in the families left behind, but at the same time migration has been an empowering experience for women, including the transfer of information and know-how and the promotion of entrepreneurship they might apply when and if they go back at home. Women's migration is important also for the families they leave behind, which may increase their income through "social remittances" and the possibility to access education and health services.

In respect to the third element, the broad availability of ICTs is also a distinctive characteristic of the current migrations. In fact migrants demonstrate a use of ICTs, such as Web 2.0, mobile phone and satellite television, which is often stronger than that of natives. Migrants can use the new media for mitigating the trauma of separation, being more able to keep in touch among themselves and with those left in their country of origin, but also for handling their life more easily in the new communities. Migrations are a condition for understanding the space and for making sense of it, but the availability of ICTs transforms the perception of physical distance and social isolation as well as the elaboration of the sense of belonging. Migrants are at the intersection of a "bonding" and a "bridging" use of ICTs, to borrow Putnam's expression (2000). However, they can also use the new media to restrict themselves within their own culture, thus paying less attention to learning a new language and culture.

Finally, as to the fourth point, migrants are different today because their condition of life is in tune with the high level of mobility that host societies have developed. A large number of their citizens are tourists, travelers and commuters. They directly know how one feels when he/she is a "stranger". Nowadays locals and migrants are individuals who both have experienced different degrees of mobility. Modern migration and other contemporary displacements such as refugees and human trafficking, but also commuting, tourism and traveling, despite their significant differences, can be seen as variations of an old human predisposition and practice (Urry 1990). At the same time, each of these social phenomena resonates with the others and intertwines reciprocally. The world is no longer unknown as before: tourists, travelers and commuters—in addition to the media—create collective knowledge, information and experience at the social level, on which also migrants can, in some ways, build upon. The sense of a global world that these new possibilities of communication and mobility generate serves to mitigate and transform the impact of migration on existing social structures. Also, the classical "chain" of migrants now acquires a more sophisticated dimension of virtual and real networks with a variety of forms. These considerations also show the necessity to historicize the migration phenomenon. Today migrations are different from the past and if we want to explore them we should contextualize them in the late modernity that is in a world shaped by a new level of agency and power by multitudes and a new level of command and domination by power elites on societies. Our aim in this collection is to precisely examine this human practice in the context of contemporary societies and post-modern media.

In a global age, all of us, including migrants, participate in multiple socio-cultural worlds and contexts. Enjoying different cultures and lifestyles has become emblematic of the post-modern condition. For migrants, their adapted and original homes constitute part of a globalized homeland. They watch local television as well as overseas channels featuring their places of origin and share national concerns while maintaining their ethnic interests. These antinomies that characterize migrants' lives are often not lived in a dramatic way, but rather they are the acquired new threads with which they weave a richer life. Also natives in the new global urban environments deal with different cultures in various ways, which go from enthusiastic appreciation to indifference and to hostility. However, migrations are not a monolithic phenomenon: there are migrants who, as we said, enthusiastically adhere to the lifestyle, social values, daily routines and identity of the host country because they are attracted and fascinated by it; there are others who do not like the host culture but who are obliged to accept and adopt it because this country gives to them the opportunity to work, to have an income, to have more civil rights and so on; others try to negotiate their own culture with that of the host country, producing a sort of cultural hybridization; and finally others who devaluate the culture of the hosting country and pass their life criticizing and stressing the limits of the new way of life they are trying out.

A Reflection on Diasporas

A large part of studies on diasporas, and also some included in this volume, focuses on the social networks built by the migrants belonging to the same culture. In this concern, a supplementary sociological reflection is useful in order to understand diasporas as social groups and the reasons of their peculiar logic. We turn again now to Simmel (1998, 568–570) to help us in understanding this point. Although, as reported above, contemporary societies generally satisfy their increasing need for differences through the development both of social and personal differentiation and of mobility, in some cases differentiation and mobility diverge. This happens when (following Simmel) stable societies differentiate internally, while migrant groups become uniform. Those who migrate depend mostly upon their fellow migrants and have more common interests with them in comparison with sedentary groups in the host society. These basic interests become stronger than the individual multiplicities and possible contrasts. For Simmel, migrating implies individualization and isolation as it removes from individuals the support of their hometown. Instead, it obliges individuals to rely exclusively on themselves, and in so doing it pushes them into a tight grouping which makes the "normal" societal differences less important.

Myths, Metaphors, Symbols and Emotions Connected to Migrations

In the social representations of migrations, myths and metaphors play an important role. Central to this are the myths of Eden or of ancestral origins as well as that of the holy land. The classical world has elaborated the myth of traveling from the reign of the living to that of the dead such as in the epopee of Gilgamesh, the myth of Orpheus who goes down into Hades and snatches Eurydice out of death's hands, the wonders of travel lived by Odysseus and Enea and Dante's travel into hell, purgatory and heaven in the Divine Comedy (Gasparini 2000, 38). Asia too has similar stories of heroes overcoming great odds such as the Monkey God transmitting Buddhist culture across the Himalayas to China, Korea and Japan. These myths serve to weave the imagination needed to leave behind one's birthplace with its particular protective energy (the genius loci) and often resulting in a harmonious (integrated) life-mode. Memories of one's homeland generate the symbolic patrimony that nourishes nostalgia, which is probably one of the most important moods migrants experience in their new country. These sojourns in foreign lands often combine great sacrifices as well as delivering unexpected rewards. This theme of transhumance vs. permanent settlements has found different expressions throughout human history from the wanderings of pastoralists to the great cities of the old world. A powerful metaphor in this regard is one that relates migration to the displacement in space with the progression of time for human beings. Each journey, in fact, is considered a paradigm of life or a metaphor of the human condition (Widman 1999). The departure is equalized to birth, the travel through life and the arrival at death. Taken together these mythological and metaphorical aspects of migrations are not only coupled with emotions like nostalgia but also with hope and curiosity, which are powerful forces that motivate people to travel around the world.

The Metamorphoses of Migrants' Identity

Migration is a basic phenomenon of the whole human history, but it is also a social process, which involves spatial dynamics and is characterized by a temporal structure. In the new country migrants sometimes become sedentary while at other times they continue to be mobile in their new territory. Migrants generally do not remain migrants all their life, and may well acquire a new citizenship from the host country. Furthermore, they may become fathers and mothers of children who are considered native, being born in the host country. So being a migrant is not a dimension which is automatically inherited as such by migrants' children, who are often at the intersection of two different cultures and have to mediate not only for themselves but also for their parents. It is a tension *en plus* that migrant children and adolescents have to cope with in addition to the tensions typical of their age group (see Lelia Green and Nahid Kabir's chapter in this volume). The multifarious dimension of migration becomes still more complex in the light of racial and ethnic group intermarriages. In several countries the marriage with a native citizen closes down a story of migration, because it is the means to obtain citizenship of the host country. Being a migrant is a transitory status and thus it creates problems of conceptualization and definition of the migrant by the various Censuses that attempt to record their presence (see Alice Robbin's chapter in this volume). Both old and new media are the main places where these tensions, these uncertainties of conceptualization and the related debates are expressed.

The Present-Absent Migrant

If we look now at what happens in the host society, we find a new general trend on the part of migrants to strengthen their self-identification. Picking up on some points from the general debate on migrations which form the background of our discourse, and which might show how much the contemporary communities of migrants are different from the past, it is sufficient to say that the typical condition of presence-absence which is experienced by people in using ICTs specifically involves migrants. Through their use of ICTs migrants are in fact absent, but somehow also present in many communities of origin—communities which started by considering migrants as part of their everyday life. This is made possible particularly because migrants are able to be in contact with their home village via the mobile phone (Nyaga Mbatia et al., n.d.). For the community of a village in Africa or in Asia, migrants are no longer people completely lost in the course of the daily routine but, for example, they are counted as if they were somehow present in the village. This "counting" as being part of the everyday life of the migrant's home village also means that new political attention is put on migrants by the country of origin; they are not seen anymore as a loss, but as an economic and political richness (as explored by Polina Stoyanova and Lilia Raycheva in their chapter in this volume).

We Are All Migrants in the Virtual Space

The negotiation between natives and immigrants regarding space is happening not only in the physical space of towns and villages but also in the virtual space designed by online communities. Online, migrants' communities share with any other communities the space at the same level of power: this is in the sense that all of us—native and migrants—migrated in the virtual space, as Castronova states in his last volume (2008). The Internet equalizes natives and migrants in offering to both the same experience of virtual migration. Thus the Internet too has contributed to make the slogan "we are all migrants" realistic, because all of us in the last twenty years—natives and migrants (in a literal sense, different from that used by Prensky 2001)—have experienced at least the kind of migration that embodies the experience of our virtual displacement in the Web. For migrants, the migration in the virtual space assumes the dimension of a strengthening of a dimension already experienced in the real world and repeated also in the virtual one. This role of the Internet as a powerful tool of diffusion of the migration experience makes it increasingly necessary to conceptualize further the new and the old media used by migrants and natives as socio-technical systems and socio-cultural environments—which both use in their daily life.

Migrants Face Communication Issues

How and when do migrants feel the need, the desire or the obligation to communicate with the purpose of managing their relational sphere? What situations create these needs, desires or obligations for communication and information? Which difficulties do migrants experience in communicating with their dearly loved ones at home, while maybe they are forgetting their mother tongue or experiencing the impossibility to explain their new life, not knowing where to start from? Migrants in fact often experience anomie in their communicative relationships with those at home as well as with the host community and groups, because they no longer share the flow and ritual of daily life with their interlocutors. What difficulties do migrants experience when they have to contract their rights in situations in which they are not able to master even the technology of language—neither orally nor, especially, written? Which sense of impotence do they feel for this communicative barrier, given that communication is the first field in which we negotiate our identity and the elaboration of reality? However, to express themselves and to keep alive the relationship with their homeland, migrants today can more often rely on a higher level of education and on more easily accessible and sophisticated means of information and communication than in the past.

Migrants, ICTs and Volatile Memory

An example of this topic is the use of online newspapers, which give the opportunity for migrants to be informed about events in their villages and towns of origin. Their forums, blogs, Web sites and social networks

contribute to create online communities based on the sharing and reinventing of a common national identity and culture (see Heike Mónika
Greschke's chapter in this volume). Virtual communities in cyberspace
allow for old and new identities to merge and coalesce. Cable networks and
global entertainment ensure that traditional as well as contemporary narratives of the self and storytelling are readily represented and performed.
Landlines, mobile phones and voice over the Internet allow for different
types of electronic communication, enabling distinct identities and intimacies (Baldassar 2007). Mediated communication (Fortunati 2005) and electronic emotion (Vincent and Fortunati 2009) are now primary supports of
social relationship and contribute to the creation of social structure. Long-
distance relationships are now a common feature of life and most people
have adopted appropriate strategies for dealing with absent others. Telecommunication, which is also tele-absence as Manovich stresses (2001), is
often immediate and oral and does not automatically leave any memories or
traces (despite the presence of recording devices). In the past, many migrant
families were able to keep track of their history through family letters and
other correspondence. In the current broadband societies the preservation
of memory becomes more difficult. They are the products of a life on the
run, where activities are now recorded as they take place (e.g. Twitter), but
are not considered worthy to become part of a lasting personal and social
memory. Whether this kind of electronic communication can reconstruct
identity narratives and family histories is as yet uncertain. Everything is
volatile, as one of our informants said, and nothing remains recorded.[3] Perhaps this is a consequence of post-bourgeois life where, as suggested in the
works of Marx and Engels, everything solid melts into air. Ironically, post-
modern life has reverted to a form of pre-literate orality or to electronic
forms of writings which are constantly deleted (Green 2009). This preference for the ephemeral, momentary and immediate fits into contemporary
cultures with their excess of meaning but lack of sense. As a response to
this surplus of meaning, the original homeland is often romanticized as
providing a more secure anchorage in a former life.

Mediated Migrations, Communicative Environment and Polymedia

The new media, and in particular the Internet, also offer written and
desynchronized forms of communication: instant messages, chats, e-mails
and social networks. In addition to memory volatility, many of these written communications might resemble oral exchanges rather than literary
texts. Initially, people feared the loss of literary skills in these new forms of
communication (new spellings, abbreviations, acronyms) but soon realized
that they constitute new ways of expression appropriate to the medium
of communication (Baron 2008). The new media however are important
communicative environments because they have their own particular transformative potential and so they contribute, along with the old media, to

co-construct a different ecosystem of migrations. In 1998 Fortunati introduced the concept of "communicative environment" to designate the place where body-to-body forms of communicative sociability coexist and interplay with the use of a more or less large spectrum of media. In 2011 Madianau and Miller proposed the notion of Polymedia to express the broad range of media which are at the disposal of individuals nowadays. Media preference by migrants and natives is highly informative. What communication technologies do migrants, their families and friends use, prefer or experience: the old, the new or a combination of the two? Does the opportunity provided by broadband technology enhance the communication and information process or does it simply inundate and overwhelm its users?

Material and Immaterial Labor is Moving

The present stage of globalization involves poor people, elites, refugees, traffickers, intellectuals, students, researchers and managers. So now the social stratification of migrations reflects more that of the society of origin. Workers now are accompanied by students, professors, researchers, managers, consultants, functionaries, travelers. It is very likely that the uses of ICTs differ for each category and one needs to know more about these differences. With the alleged weakening of national identity in the context of globalization, to what extent do migrants still form their original communities or are they more willing to adopt new mentalities and forms of living in the host countries? In addition to the diffusion of ICTs, powerful social processes such as individualization with the consequent effect of the weakening of families' relationships and the transformation of family roles generate new imperatives for migrants. Another consequence is the current trend of self-identification by migrants negotiated on the basis of the behaviors of their communities in their countries of origin. The repression of Uyghurs (who are Muslims) in China in July 2009 immediately unleashed many tensions, for example, in Algiers between the local community (which is Muslim) and the Chinese one.[4]

Migrations and Generations

Last but by no means least, generation plays a particular role in recent migrations. A large part of the literature on migrations focuses on first generation migrants (see the chapter of Clifton Evers and Gerard Goggin in this volume). But second generation migrants also deserve the same attention, as shown in the chapter of Green and Kabir in this volume. Some concepts which might frame the theoretical background of the studies on this topic come again from classics. It must be acknowledged that migrations studies generally use one particular notion of generation—the genealogical one—because this has a high information value, particularly when the individual of reference undergoes a traumatic change as it is in the case of migrants (Fortunati 2011, 202). In fact, the second generation of migrants creates serious problems for the uniformity of migrants' groups, of which we talked above, essentially for three reasons: first, new generations are not repositories of societal norms but are subjects that challenge existing societal norms and values and bring social effervescence (Mannheim 1952); second, the need of refusing parental authority and of identifying with one's own peers (that is, the need to identify themselves as children and as young) is stronger than the need to identify as migrant; third, the language they speak is generally that of the host society and by speaking this language they also pick up the culture. Only the study of the complex interplay of migrations and generations allows us to understand why the second generation of migrants shows a much higher degree of dynamic dealing with the host culture.

OVERVIEW OF THIS VOLUME

This book covers five theoretical and empirical areas of research: 1) conceptual tools to analyze migrations; 2) migration as a factor of gender and generation transformative processes; 3) new media as a lens to look at migrations and diasporas in a new way; 4) religion, mobility and social policies as shaped by migrations; 5) a case study: China and its internal migrations and diasporas as well as expatriates in China. Using the concept of migrants as e-actors (digital subjects) (Fortunati et al. 2010), this book is meant to conceptualize the complex and crucial role migrants play in the information society and the enormous potentialities they represent for the development of a knowledge society.

The first selection of chapters "Conceptual tools to analyze migrations" discusses how migrants are concerned in their everyday life with different notions, models, approaches and languages and explores them with three specific case studies from Italy, the U.K. and U.S. The authors debate and challenge different ways to define and conceptualize migrations, different strategies and methodologies to investigate this phenomenon at the social level and finally, different approaches and perspectives to analyze the first technology that migrants have to domesticate—their language. In other words the chapters in this section deal with the development and promotion of an anthropocentric perspective in the dealing of migrants with the old and the new media. The fascinating and passionate reasons of the intercultural perspective are discussed and expanded by Giuseppe Mantovani, who argues for the importance of this approach for migration studies. The interplay between elements such race, ethnicity and culture is one of the fundamental concerns of cultural theory. Second, the socio-linguistic problem of the migration of languages, with special emphasis on English, which has become in the course of time Englishes, is illustrated by Maria Bortoluzzi. Finally, Alice Robbin discusses the way in which the debate, recently reported by media, has framed new categories such as "multiracial". In the cartography of races and ethnicities, the category multiracial is analyzed as a specific case study, strengthened recently by the election of President Obama. This section emphasizes the role of the migrant as e-actor and the way that this symbiosis of humans and their electronic interface with ICTs characterizes migrants' participation in 21st century broadband society. The increasingly high penetration of broadband conveys inevitable tensions between migrants and natives which are explored within these chapters with the emphasis being on the push from the user rather than the pull from the technology.

The second section, entitled "Gender and generation intertwining with migrations", deals with how migrations often carry out significant and rapid changes in the way people construct their gender narratives and interact with the social construction of femininity and masculinity in the host countries through the old and the new media. Three specific case studies, one from Jamaica and two from Australia, are examined. Heather Horst discusses this point from the particular perspective of how Jamaican transnational communication is characterized by a gendered dynamics. The following chapter written by Clifton Evers and Gerard Goggin reconstructs how mobile communication plays an important role in the negotiation of new cultural dynamics by young men living in suburbs in inner and outer Sydney. Migration often redesigns the structure of power, the roles played inside the family, as well as the way in which parents have to redefine the educational and caring processes of the new generations. To conclude this section, Lelia Green and Nahid Kabir illustrate their research project on a group of adolescents in Australia with a double national identity. These subjects show how their respondents negotiate their diasporic identity at the same time as they struggle for their own personal autonomy distinct from family and community.

The third section "Looking at the migrations and diasporas through the lens of the new media" looks specifically at why the emergence of new forms of agency, subjectivity and mediated communication are an important aspect of developing broadband societies. Their role in social interaction in public and private spaces is increasingly important for the users of all types of ICTs, but most particularly the mobile phone. These three chapters examine the complexity of the weave of communication in everyday life in the context of digitally mediated technologies with three case studies from Philippines, Paraguay and Bulgaria. Raul Pertierra opens this section with his chapter on the Filipino diasporas, stressing how the mobile phone, the Internet and Web 2.0 are used by Filipino migrants to maintain and reproduce original relationships but also to acquire new communicative and social dimensions and possibilities both abroad and at home. The following chapter by Heike Mónika Greschke analyses how the Internet allows migrants to maintain links between sites of being and sites of belonging and to reshape fundamental notions such as homeland and the sense of proximity and togetherness. This section concludes with the chapter of Polina Stoyanova and Lilia Raycheva, who discuss important issues of the Bulgarian Diasporas by analyzing the diaspora of the Bulgarian media. In this chapter the authors investigate and evaluate the role of the information and communication technologies for social inclusion within the local media consumer community, as well as for the process of national "integration". New psychological, emotional and affective dimensions are produced by migrants in virtual communities or as audiences, who might have the possibility to continue to speak their own language if, as in the Bulgarian case, media have also experienced a kind of diasporic process in order to be near and to support Bulgarian migrants all around the world.

In the fourth section "Religion, mobility and social policies: How migrants' use of the new media is shaping society" the discourse is focused on how migrants' use of ICTs influences social phenomena such as religion, identity, social mobility and social policies. Migrants deal with agency, subjectivity and identity as a result of the new mediated interpersonal communication in the broadband society. These new identities and subjectivities play significant roles for the reorganization of social practices, forms of ritualization and structures with case studies from Congo and five European countries: France, Germany, Italy, Spain and the United Kingdom. The way migrants manage their electronic identities and access ICTs in public and private spaces is crucial for the transformations of their everyday life, signified by different practices, routines and meanings also at a religious level as explored by David Garbin and Manuel A. Vásquez in their chapter. This section maps the emerging research fields relevant for understanding uses and relationships with ICTs (e.g. cognitive complexity, group decision-making, ethical aspects, etc.). It also explores to what extent migrants are mobile in the new countries with the purpose of verifying whether the relationship between migrants and mobility is invariant or episodic as in the chapter by Andreas Hepp, Cigdem Bozdag and Laura Suna. In the last chapter of this section Stefano Kluzer and Cristiano Codagnone discuss why nowhere in Europe can there be found clear policies and/or strategies that address ICT opportunities for immigrants and ethnic minorities.

Finally, the fifth section illustrates a powerful case study: "China and its internal migrations, diasporas and expatriates". The economic revival in China is best represented by its employment of the new media such as the Internet, blogging and mobile phones. In China, by the middle of 2010,[5] there were 420 million Internet connections and in September of the same year there were 833.3 million mobile phones.[6] The consequence of this high volume of users is that the numbers of Chinese who use the mobile phone and who go online are both higher than their American counterparts. Several types of migrations are studied with regards to China with the purpose of understanding how the use of the new media serves to construct the contemporary Chinese identity. They comprise internal migrants moving from the inner regions of China to the coastal regions of the south; Chinese migrants moving to Australia and

Italy and expatriates from several countries now living in China. This leads to differing dynamics of identity and tactics of co-construction of a new culture. China is emerging as among the most important of the research propositions in this volume, not just because of its massive population but also for the new cartographies of ICT uses. This section uses ICTs as the focus for the exploration of the social processes of negotiation that currently take place both inside internal migrations in and from China as well as migrations to China. Pui-lam Law as well as Chung-tai Cheng depict how Chinese migrants arriving from internal regions to the towns of the industrialized coasts take advantage of the mobile phone to negotiate a more acceptable life. They also observe that this use in several cases can easily transform itself from a moment of relaxation to relieve the stresses of long hours of work in heavy alienation and in lengthening the working day. Tom Denison and Graeme Johanson, on the other hand, have conducted research on Chinese diasporas, in particular migrants in two different settings, Prato, Italy and Melbourne, Australia, that investigates the role of ICTs in contributing to social cohesion and in maintaining social networks. Finally, David Herold focuses on expatriates in China as he investigates their opinions and attitudes reported in many blogs. Herold tells us that there are 150,000 foreigners legally working in China (frequent business travelers, students, spouses, etc.) whose influence on the opinions of, e.g., foreign journalists publishing news stories in European and American media outlets, is not negligible.

NOTES

1. http://www.iom.int/jahia/jsp/index.jsp (accessed April 20, 2011); United Nations, Department of Economic and Social Affairs, Population Division (2009). *Trends in International Migrant Stock: The 2008 Revision* (United Nations database, POP/DB/MIG/Stock/Rev.2008) http://esa.un.org/migration/p2k0data.asp (accessed April 20, 2011).
2. Among them, 20 to 30 million are unauthorized, almost 16 million are refugees.
3. This concept was elaborated by a woman living in Madrid interviewed by Leopoldina Fortunati three years ago. She belonged to an Italian family that emigrated to Argentina at the beginning of the 20th century and she was mother of a daughter living in Africa. In the interview she compared the different effects that analogical and digital cultures had produced in her family.
4. http://en.wikipedia.org/wiki/July_2009_%C3%9Cr%C3%BCmqi_riots
5. http://research.cnnic.cn/html/1279173730d2350.html
6. http://www.miit.gov.cn/n11293472/n11293832/n11294132/n12858447/13451760.html

REFERENCES

Anderson, Benedict. 1983. *Imagined communities: Reflections on the origin and spread of nationalism.* London and New York: Verso.

Appadurai, Arjun. 1996. *Modernity at large: Cultural dimensions of globalization*. Minneapolis: University of Minnesota Press.

Baldassar, Loretta. 2007. Transnational families and aged care: The mobility of care and the migrancy of ageing. *Journal of Ethnic and Migration Studies* 33(2):275–297.

Baron, Naomi Susan. 2008. *Always on: Language in an online and mobile world*. Oxford: Oxford University Press.

Castronova, Edward. 2008. *Exodus to the virtual world: How online fun is changing reality*. New York: Palgrave MacMillan.

Corsi, Dinora, ed. 1999. *Altrove: Viaggi di donne dall'antichità al novecento [Elsewhere: Women's travels from the antiquities to the twentieth century]*. Rome: Viella.

Donner, Johnathan. 2008. Research approaches to mobile use in the developing world: A review of the literature. *The Information Society* 24(3):140–159.

Dustmann, Christian. 2005a. Gender and ethnicity: Married immigrants in Britain (with Francesca Fabbri). *Oxford Review of Economic Policy* 21:462–484.

———. 2005b. The assessment: Gender and the life cycle. *Oxford Review of Economic Policy* 21:325–339.

———. 2008. Intergenerational mobility and return migration: Comparing sons of foreign and native born fathers. Discussion Paper 05/05, Centre for Research and Analysis of Migration (CReAM), University College London, Department of Economics.

Ferrarotti, Franco. 1999. *Partire, tornare: Viaggiatori e pellegrini alla fine del millennio [Leaving, coming back: Travelers and pelerines at the end of the millennium]*. Rome: Donzelli.

Fortunati, Leopoldina, ed. 1998. Telecommunicando in Europe. Milana: Angeli.

———. 2005. Is body-to-body communication still the prototype? *The Information Society* 21(1):53–62.

———. 2011. General native generations and the new media. In *Broadband society and generational changes*, ed. Fausto Colombo and Leopoldina Fortunati, 201–220. Berlin: Peter Lang.

Fortunati, Leopoldina, Jane Vincent, Julian Gebhardt, Andraž Petrovčič, and Olga Vershinskaya, eds. 2010. *Interacting with broadband society*. Berlin: Peter Lang.

Gasparini, Giovanni, ed. 2000. *Il viaggio [The travel]*. Rome: Edizioni Lavoro.

Goggin, Gerard. 2008. Reorienting the mobile: Australasian imaginaries. *The Information Society* 24:171–181.

Green, Nicola. 2009. Mobility, memory, and identity. In *Mobile technologies: From telecommunications to media*, ed. Gerard Goggin and Larissa Hjorth, 266–282. New York: Routledge.

Heckmann, Friedrich, and Dominique Schnapper. 2003. *The integration of immigrants in European societies: National differences and trends of convergence*. Stuttgart, Germany: Lucius and Lucius.

Hochschild, Arlie Russell, and Barbara Ehrenreich, eds. 2003. *Global woman: Nannies, maids and sex workers in the new economy*. New York: Metropolitan Press.

Madianou, Mirca, and Daniel Miller. 2011. *Technologies of love: Migration and the Polymedia revolution*. London: Routledge.

Mannheim, Karl. 1952. The problem of generations. In *Essays on the Sociology of Knowledge by Karl Mannheim*, ed. Paul Keeskemeti, 276–320. London: Routledge and Kegan Paul.

Manovich, Lev. 2001. *The language of new media*. Cambridge, MA: MIT Press.

McLean, Iain. 1996. *Concise dictionary of politics*. Oxford: Oxford University Press.

Miller, Daniel, and Heather Horst. 2006. *The cell phone: An anthropology of communication*. Oxford: Berg.

Nyaga Mbatia, P., A. Palackal, D-B. S. Dzorgbo, R. B. Duque, M. A. Ynalvez, and W. Shrum.

n.d. Mobile Telephony and Core Network Expansion in Kenya. Unpublished paper.

Pertierra, Raul. 2006. *Transforming technologies: Altered selves*. Manila: De La Salle University Press.

Pingol, Alicia. 2001. *Remaking masculinities*. Quezon City: University of the Philippines, Center for Women's Studies.

Prensky, Marc. 2001. Digital natives, digital immigrants. *On the Horizon* 9(5). MCB University Press. http://pirate.shu.edu/~deyrupma/digital%20immigrants, %20part%20I.pdf (accessed December 20, 2009).

Putnam, Robert D. 2000. *Bowling alone: The collapse and revival of American community*. New York: Simon and Schuster.

Scott, John, and Gordon Marshall. 2005. *Dictionary of sociology*. Oxford: Oxford University Press.

Silvestre, Maria Luisa, and Adriana Valerio. 1999. *Donne in viaggio: Viaggio religioso, politico, metaforico*. Rome and Bari: Laterza.

Simmel, Georg. 1998. *Soziologie: Untersuchungen über die Formen der Vergesellschaftung*. Trans: *Sociologia*, Turin: Edizioni di Comunità.

Sombart, Werner. 1902/1927. *Der moderne Kapitalismus*. München and Leipzig: Duncker and Humblot.

Urry, John. 1990. *The tourist gaze: Leisure and travel in contemporary societies*. London: Sage.

Vincent, Jane, and Leopoldina Fortunati, eds. 2009. *Electronic emotion: The mediation of emotion via information and communication technologies*. Oxford: Peter Lang.

Widmann, Claudio, ed. 1999. *Il viaggio come metafora dell'esistenza* [*The travel as metaphor of the existence*]. Rome: Edizioni Scientifiche Ma.Gi.

Theme 1

Conceptual Perspectives of Migrants in Post-Modern Societies

1 New Media, Migrations and Culture
From Multi- to Interculture

Giuseppe Mantovani

INTRODUCTION

The new media are socio-cultural environments in which people live their everyday experiences, globally and locally situated (Fortunati 2005). They offer not only new ways to connect with other people but also spaces for projects, feelings and imaginations unthinkable before. The approach of cultural psychology used in this chapter provides a strong foundation for exploring this topic based on the concepts of mediation and artifacts. For cultural psychology the advent of the new communication technologies (artifacts) affects, of necessity, the ways people experience reality, create its structure and find its internal boundaries. Using new media in everyday life has enlarged and somehow complicated the ways people have of making sense of the situations in which they find themselves (Mantovani 1996a, 1996b, 2002). Furthermore, the contribution that cultural psychology brings to research on new media is clarified through comparison with the way cross-cultural research views cultures and cultural differences (Berry 1997, 2001).

The concept of culture is critical to understanding migration processes in which different cultural worlds experience close contact. Further, the meaning of concepts such as community, identity and transnational depend on the framework (multicultural or intercultural) that is adopted. The multicultural approach is built on an idea of culture as a distinctive property of a social group. This is a community, internally homogeneous and separated from the others, whose members share a fixed identity assigned to them by tradition. The intercultural approach, on the contrary, embodies a narrative, pluralistic, open concept of culture: borders of every kind are continuously crossed by people with different backgrounds in interchanges that mix commodities as well as experiences, ideas and imaginations. The development of the new media has been, and still is, the strongest support for the growth of intercultural processes; it is not even possible to draw a sharp separation between the development of the new media and that of the current intercultural exchanges (Castells 1996). Using the concept of diaspora (Appadurai 1998; Pertierra, this volume) implies the adoption of an intercultural view of migration focused on mobility rather than on stability. In the following pages the differences between multiculture and interculture are presented from the situated point of view of a social psychologist committed to the intercultural perspective.

TWO ALTERNATIVE CONCEPTIONS OF "CULTURE": "REIFIED" AND "NARRATIVE"

The first conception, that of reified culture, is centered on social groups and their identity. Culture is constructed as an identity marker, something that both identifies and separates a group from the other groups. Culture is an objective property of a group, a badge that makes manifest to which group a person belongs and separates its members from non-members on the basis of a set of objective characteristics. Metaphors such as heritage, roots and traditions are often used in association with the central concept of identity. Identity comes from inside the group, unaffected by the relationships it can have with other groups. Consider for instance the genealogy of Europe (the traditional basis of Western identity) thought to be born in Athens (and later, partially, in Rome) in isolation from the surrounding societies. Groups are internally homogeneous and neatly separated from the other groups. Borders among groups are strictly controlled to defend the purity of the group (its orthodoxy, its heritage and so on). Baumann (1996, 1999) calls this conception reified because it treats culture as a datum, what one might refer to as a "thing-in-the-world-out-there", something that requires only to be acknowledged. A member of a culture has only to register its presence as something already-existing-out-there.

The second and alternative conception, narrative culture, is centered on people's agency: people are not cultural clones but active, creative, and fully responsible social actors. Culture is not a badge, a property, a thing that people share with the other members of their groups, but a social construction, a concept, a tool—used mainly by outsiders—to organize, administer and/or categorize aspects of a given society. Culture is a concept used to address some social processes; it is not a super-individual independent agent dictating people's everyday choices. Culture in this second sense has been presented in various and converging ways: as a mediation device (Cole 1996), as a set of resources for action (Mantovani 2000, 2002, 2004a), as a narration "shared, contested, and negotiated" (Benhabib 2002, 1). This image emphasizes—in contrast with the first conception of culture—the plurivocal, polyphonic, pluralistic character of culture which refers to a social construction not only (somehow) shared but also contested (as every social actor has peculiar circumstances, goals, resources) and negotiated. Negotiation is not an occasional but a basic activity among humans. Groups are neither homogeneous nor separated, as the first conception of culture imagined: borders are always more or less porous and people continuously cross them, sometimes even live on them. Identities do not grow in isolation but are born through enduring patterns of interactions, e.g.

early European culture and identity incorporated important elements from Egypt, Persia, Phoenicia and so on; the Hebrew parenting of early Christianity and the powerful Islamic influence on many aspects of European life in past centuries are further instances of the intense exchanges taking place across borders.

THE EXAMPLE OF CONSTRUCTING "SECOND GENERATION IMMIGRANTS"

Let me now consider the differences between the two conceptions explained in the last section through an example, namely the ways in which so called "second generation immigrants" are categorized by researchers, social operators and policy makers. The differences between these two approaches are apparent in the way the borders separating (for the first) or connecting (for the second) cultures are seen. In the first case borders are assumed as impregnable and the characteristics of people crossing borders are minimized and stigmatized as belonging neither to one nor to the other community. In the second case borders are seen as porous and people crossing them are considered a potentially precious resource for themselves and for the social groups to which they belong. The way borders are treated summarizes the way cultures are considered.

It is curious, as Baumann (1996) asserts, that persons born in the country who are therefore legal citizens of it can currently be labeled as immigrants, although they are at least one generation removed from the first point of immigration. In the U.K. the label most often applies only to the children of families from Asia, Africa or China—few would call a person born in the U.K. of a family of German or Swedish origin a second generation immigrant. Nevertheless second generation immigrants are judged at risk because they are seen as suspended between two cultures. Why, asks Baumann, should these persons be precariously suspended between rather than belonging to two (or more) cultures? The choice between the two visions depends on the conception of culture one adopts. If the reification concept is accepted—and groups are conceived of as homogeneous and separate—the very idea of people belonging to more than one group will appear improper. The situation of a person who is at the same time seen to be a U.K. citizen and a Muslim from Pakistan may be judged by some differently from a person who has indigenous U.K. origins. The second generation immigrants, in this perspective, are not accepted by some as true, pure, authentic British citizens simply because of the origin, religion, color of the skin and so on of their parents. At the same time they are not accepted as true, pure, authentic Muslims from Pakistan, or Sikhs from East Africa, or West Indians from the Caribbean because of their U.K. birth and education. Thus for this first conception of culture the second generation immigrants are persons internally divided and socially problematic, a potential threat for themselves, their relatives and both their country of birth as well and their parents' country.

For the narrative interpretation of culture the second generation immigrants are not a problem at all. On the contrary, they are considered a precious resource enabling the persons involved to speak more than one language (language is the meta-artifact that supports every cultural system), to understand multiple voices and to participate in different histories. The double (or triple) heritage of the second generation immigrants can be fortuitous not only for the individuals but also for the societies in which they live: the more permeable the borders are, the more a society is ready to participate in the ongoing global world negotiations. The idea of culture proposed about seventy years ago by Michael Baktin, a Russian critic of literature, illuminates the global landscape which is opening before our eyes:

> One must not imagine the realm of culture as some sort of spatial whole having boundaries but also having internal territory. The realm of culture has no internal territory: it is entirely distributed along the boundaries. Boundaries pass everywhere, through its every aspect. Every cultural act lives essentially on the boundaries: in this is its seriousness and significance. Abstracted from boundaries it loses its soil, it becomes empty, arrogant, it degenerates and dies. (1981, 61)

For this second concept of culture, the real risk that second generation immigrants meet because of their origins is that they are in danger of becoming victims of the fundamentalist vision of culture that depicts them as a potential threat to themselves and to their communities.

CONCEIVING CULTURAL DIFFERENCES AS ONLY QUANTITATIVE: CROSS-CULTURAL RESEARCH

There appears to be little agreement on how cultural differences can be understood. One position affirms that differences among cultures are only quantitative while the other states that differences can be also qualitative. According to the quantitative approach cultures are perfectly comparable; differences consist only in the scores marked by members of different societies when responding to the same standard items. In contrast, the qualitative approach thinks of cultures as incomparable: every language resists perfect translation; every social structure is unique and emotions and cognitions produced in a given culture are impenetrable by persons living in other cultures. The first of these two positions is explicitly assumed as the basis for cross-cultural research while the second supports the recent advancements in cultural psychology (Cole 1995, 1996; Cole, Gay, and Glick 1971; Shweder 1991, 2003). Cross-cultural research assumes that cultures have comparable structures (homologues) and are made by the same elements;

this is why cultural differences can be studied by comparing the results obtained from the application of standard tools (usually items of Likert scales (1932) for the study of attitudes) to members of different societies. Typical cross-cultural research findings are, for example, that members of traditional societies prefer a collectivist Self while U.S. citizens favor an individualist Self (Triandis 1989), Japanese people develop an interdependent Self while U.S. citizens develop an independent Self (Markus and Kitayama 1991) and so on. Furthermore, cross-cultural research tends to consider Western societies as a homogeneous block, comparing them with the similarly homogeneous block of Oriental or Japanese or traditional (i.e. non-Western) cultures (and ignoring in both instances what historians, anthropologists, linguists and even political scientists would say about differences within the imaginary Western and non-Western blocks).

The theory and methodology embodied in cross-cultural research are also open to criticism on various points. Exploring theory first: the difficulties created by the colonial opposition of "the West versus the Rest" of the world conveys a strong ethnocentric prejudice about Western superiority (Mantovani 2004a). Contrasting the West with the Rest (or with particular areas of the Rest) is a trait of cross-cultural research that can be traced from the first cross-cultural field study, the expedition promoted by the University of Cambridge in 1895 (Jahoda 1992) to the Torres Straits, dividing Australia from New Guinea, in which natives' eyesight was compared with Europeans'.[1] A further theoretical exemplar concerns the use of very large categories devoid of precise analytical references such as Western, non-Western, traditional, Orientals, Chinese and so on, which infringes the deontology norms that forbid psychologists to promote stereotypes that can hurt people of non-Western societies because of the ignorance of the peculiar characteristics of their societies. For example Chinese and Japanese, Hindu Indians and Muslim Pakistani, Balinese and Kashmiri may consider it offensive to be amassed under the dismissive label of "Asians".

I turn now to the use of standard questionnaires that categorize and measure different cultural contexts (usually Likert attitude scales). The presumption here is that not only the terms used in the questions (and in the answers given by the participants to the cross-cultural research) but also the general sense that participants can make of the question-answer situation will be exactly the same in different cultural contexts (this is the core of the application of a questionnaire: see Schwarz 1999). This seems hardly acceptable to linguists, conversation analysts, cultural anthropologists and social psychologists sensitive to the role of context (Duranti 1994, 1997; Duranti and Goodwin 1992). Another methodological concern regards the approach that establishes a global and massive comparison between (members of) different cultures such as, for example, Americans and Japanese. This is open to criticism for at least two reasons: First, the persons who respond to the questionnaires are neither representative of large categories such as Americans or Japanese (setting apart the fact that the source of these psychological data are often college students) nor of sub-categories such as African American; Irish American; Japanese American and so on. Second, as is apparent from the last point, the categorization which under-lies all the research programs of cross-cultural research adopts—explicitly or tacitly—a reified, fundamentalist conception of culture. For cross-cultural research one is, for example, American or Japanese or Chinese or Mexican, but what of a child born from a Chinese American mother and an Irish American father? He/she will not enter in the research because his/her responses to the questionnaires would blur the clear distinction on which the cross-cultural research program lies. The above child will be discarded just because he/she is non-representative of his/her group and is, in effect, suspended between two cultures.

ACKNOWLEDGING CULTURAL DIFFERENCES AS QUALITATIVE: CULTURAL PSYCHOLOGY

In addition to considering the quantitative approach favored in cross-cultural studies I now examine the qualitative approach that enables the exploration of cognition, emotion and ethics which are different in our diverse human societies. Emotions that some know well as anger are unknown in other societies; even emotions some consider to be natural and innate such as fear seem to be absent in other cultures. Emotions, it appears, have to be cultivated (Despret 2001). It is no surprise that emotions born in foreign lands such as *song, metagu, amahe, ikari* and many others cannot be properly translated; cultural anthropologists (Clif-ford 1997; Geertz 1994; Hannerz 1992, 2001) and cultural psychologists (Cole 1996; Shweder 1991, 2003) explain that in order to understand them we should have been educated in the social worlds that produced them through a rich array of cultural artifacts shaping education, reli-gion, ethics and everyday routines.

Let us consider by way of explanation Shweder's (2003) analysis of the contemporary conception of an emotion, *lajja*, prominent in the Hindu moral world. "*Lajja* is often translated by bilingual informants and dic-tionaries as 'shame', 'embarrassment', 'shyness' or 'modesty', yet every one of these translations is problematic or fatally flawed" (Shweder 2003, 156). While shame conveys a negative feeling in Western societies,

> *lajja* is something one deliberately shows or puts on display the way we might show our gratitude, loyalty, or respect. It is a state of con-sciousness that has been baptized in South Asia as a supreme virtue, especially for women, and it is routinely exhibited in everyday life, for example, every time a married woman covers her face or ducks out of a room to avoid direct affiliation with these members of her family she is supposed to avoid. (Shweder 2003, 156–157)

Furthermore *lajja* is "a general habit of respect for social hierarchy and a consciousness of one's social and public responsibilities" (Shweder 2003, 161) that reflects the Hindu moral order but is basically extraneous to Western people raised in veneration of autonomy, self-expression and personal success.

It is no wonder if (some) Hindu virtues regarding women are contested by (some) American feminists. Shweder criticizes sharply the feminist disapproval of Hindu traditional virtues:

> If anthropology is a discipline that studies differences, it is necessary that feminists devise the means to analyze and interpret differences that they find personally disturbing without distorting and thus dishonoring the objects of their study. To ignore the alternative moral goods emphasized and made manifest in family life practice in India, to presume that inner control, service, and deferred gratification amount to subordination and acceptance of oppression, to represent Hindu women in South Asia as either victims or subversives is not only to dishonor these women—it is to engage in little more than a late-twentieth-century version of cognitive and moral imperialism. (Shweder 2003, 274–275)

Shweder's argument against the resurgence of ideas of Western supremacy clearly merits attention: what is at stake in the present debate on cultural diversity is the scientific, moral and political framework through which we see other societies (but, for those who accept the agency-based concept of culture, there are neither other cultures forming compact social blocks nor stereotyped others opposed to the "us" in everyday life) (Mantovani 2004b).

Modern cultural psychology provides theoretical and methodological tools to treat cultures as qualitatively different. The pathways of cultural psychology and cross-cultural research parted during the sixties, when Mike Cole travelled to Liberia (Western Africa) to work on the mathematical abilities of the Kpelle, a population of rice farmers. Cole, then a young researcher expert in probabilistic models for mathematical learning and not at all versed in cultural theory, was quick to realize that the difficulties of the Kpelle children had little to do with mental abilities and much to do with the artifacts available in the Kpelle environment (Cole, Gay, and Glick 1971). School (Western-oriented school, actually) was the artifact missing in Kpelle society:

> When I arrived, I made a point to ask people who spent a lot of time around children about the difficulties that the local children displayed in school, especially their difficulties with mathematics. The list of intellectual difficulties that the tribal children were said to encounter was a long one. They had difficulties distinguishing between different geometrical shapes because, I was told, they experienced severe perceptual problems.

This made it virtually impossible for them to solve even simple jigsaw puzzles, and I heard several times that 'Africans can't do puzzles'. I also learned that Africans didn't know how to classify and that when faced with a choice between thinking and remembering they would resort to rote remembering, at which they were said to excel . . . These assertions— which echo more than a century of European claims about the primitive mind, although I did not realize this at the time—are based upon a deficit model of cultural variations . . . I found these generalizations difficult to credit. It is a long way from inability to do jigsaw puzzles to general perceptual incapacities. Visits to schools showed me the basis for some of these assertions. In many classrooms I saw students engaged in rote remembering. Not only were children required to recite from memory long passages of European poetry that they could not understand; they seemed firmly convinced that mathematics, too, was strictly a matter of memorizing (Cole 1996, 73)We see clearly in these lines from Cole the about turn taken by cultural psychology which has shifted the focus from supposed mental abilities to the situated interactions mediated by artifacts. This realization marked a sharp separation in the pathways of cultural psychology and cross-cultural research.

"CULTURE" AS A NARRATIVE SHARED, CONTESTED AND NEGOTIATED

By asserting that people interact with their environments through the mediation of artifacts—tools both physical and ideal produced to support people's everyday activities—cultural psychology took a route diverging from cross-cultural research. Artifacts were considered not only as individual tools but also as complex systems designed for the development of complex social activities: language first (a meta-artifact) and also school, religion and family began to populate the new landscape of cultural psychology; today the new communication media are in the foreground. Exploring the tradition of cultural psychology started from the work of Lev Vygotsky, a psychologist who did innovative research in the Soviet Union in the early decades of the past century. Following Vygotsky (1978) intelligence cannot be thought of as a property of the individual mind but is best understood as the contingent outcome of the interaction of social actors with their environments. Distributed intelligence emerged as a new approach in cognitive science (Clancey 1997; Clark 1997; Lave 1988; Suchman 1987), in social ergonomics (Hutchins 1995, 1997) and in education (Rogoff 2003). Cultural psychology can account for every kind of diversity: in institutions, in moral values, in cuisine and in women's fashion. All these are configurations of interactions mediated by different sets of artifacts. It is no surprise, therefore, that for researchers adopting the cultural psychology approach every society is different. This is a sharp contrast with the cross-cultural approach, based on extensive use of stereotypes and on the assumption that societies are self-contained wholes, homogeneous inside and separate from the outside world.

The emphasis on qualitative differences favors the development of an approach that sees culture as a narrative shared, contested and negotiated:

> In my view, all analyses of cultures, whether empirical or normative, must begin by distinguishing the standpoint of the social observer from that of the social agent. The social observer—whether an eighteen-century narrator or chronicler; a nineteen century general, linguist, or educational reformer; or a twentieth- century anthropologist, secret agent, or development worker—is the one who imposes, together with local elites, unity and coherence on cultures as observed entities. Any view of cultures as clearly delineable wholes is a view from the outside that generates coherence for the purposes of understanding and control. Participants in the culture, by contrast, experience their traditions, stories, rituals and symbols, tools, and material living conditions through shared, albeit contested and contestable, narrative accounts. From within, a culture does not appear as a whole; rather, it forms a horizon that recedes each time one approaches it. (Benhabib 2002, 5)

In this statement Benhabib is stressing the plurality of actors participating in the narrative, each of them with his/her particular version of the events.

A narrative is a search for possible meanings of experiences that, for some, often seem devoid of any sense (Ochs and Capps 2001; Ochs and Sterponi 2003). Frequently other people are involved in narrations and indeed, the concept of culture as a multivocal narration is consistent with the acknowledgment of cultural diversity in a qualitative sense. Narrations do not need to conform to pre-defined standards, on the contrary, they tell particular, situated, even contradictory stories. The reified conception of culture and the cross-cultural research program which depends on it are based on the idea that cultures create social groups which are realities in themselves and that these social groups (or cultures) are separate and compact. This approach accepts quantitative differences among cultures but is hardly prepared to acknowledge qualitative differences. Quantitative differences are not captured through standardized items, questionnaires or scales which rely on the assumption that the human cognitive and emotional processes are basically invariant across cultures, only present in different degrees in each of them.

UNIVERSAL VALUES AND QUALITATIVE DIFFERENCES: THE CASE OF "DEMOCRACY"

Multiculture as a scientific, moral and political program is founded on the reified concept of culture and supported by the theoretical and methodological approach of cross-cultural research. It ignores people's agency and accepts diversity only in the sense of quantitative differences among societies and not in the sense of multiple standards for the understanding of cultures. Interculture, however, is a new emerging scientific, moral and political program founded on the narrative concept of culture and supported by the approach of cultural psychology. It is focused on agency and people's concrete differences in everyday life situations: differences in gender, age, profession, country, religion, origin, health and more. Interculture acknowledges, without problems, the existence of qualitative differences among societies: narrations are thought of as multiple and situated.

An argument often leveled against the qualitative approach to cultural differences is that of "relativism". If we accept that societies can have different standards and that no objective order or reference scale can be found for them, what will happen to universal values and rights? In principle, we do not see any contradiction between defending universal rights and anchoring them to specific historical vicissitudes. But we have to be aware of the fact that the format in which we are used to expressing universal values is that of "our" culture. Universal values are expressed in a language that is ours, from which it is difficult to disentangle them. The experiences from which our universal rights have been born—e.g. freedom of expression, free choice in religious matters, equal rights for all citizens—belong to "our" history in ways that make them hardly separable from it.

Universal rights and values normally emerge in specific historical contexts: this is not relativism, but acknowledgment of the situated nature of values and rights. Alan Dershowitz (2004), who teaches Ethics of Law at Harvard, explains that universal rights are social creations founded on experiences of the wrong. When history produces experiences of the wrong that are similar in large parts of the world's population and when these experiences become generally known, reflected upon and condemned, the corresponding universal rights are claimed. This has been the case in the past for slavery, genocide and limitations of personal freedom. The wrongs that are experienced as such at present will in due time produce the rights that will be claimed in the future.

The awareness of possible prejudice of Western superiority should stand as a caveat to "our" formulations of universal values. A good example is provided by the position of Amatya Sen (2003, 2004), Nobel Prize winner for economics, on the issue of Americans exporting democracy to Iraq. According to Sen, democracy is not an American patented good that should be shipped to an Arab country but a value pursued by different societies in many areas of the planet in different historical periods. The core concept of democracy, according to Sen's analysis, is public discussion and freedom of expression. Various aspects of democracy can be developed in different societies, from defending religious tolerance to creating legal norms for the free expression of people's will, supported by institutions spanning from tribal meetings to formal elections. Democracy as a universal value should not be identified with (and reduced to) the forms it has taken in Western societies, elections, parliaments, free press. Although these forms can be deemed important, they are partial realizations of the universal value of democracy; in this way, we can accept both universal values and qualitative differences among societies.

TO FACE EACH OTHER ON MORE EQUAL
GROUND AS MEMBERS OF FLAWED SOCIETIES

For members of Western societies it is difficult to accept that some of their traditional beliefs and practices could be challenged by members of other cultures. Renato Rosaldo (1989), an American ethnographer, narrates an illuminating experience which occurred to him. In the late sixties, when doing field research among the Ilingot of northern Luzon, Philippines, he struggled against his revulsion towards one of the central local practices, headhunting. When he questioned his Ilingot informants about headhunting they assured him that no heads had been taken since the end of the Second World War. After staying in Luzon for almost one year, Rosaldo was accepted as a member of the Ilingot society. At this point, while flying over the island, his Ilingot brother Tukbaw, pointed down and said, "There is where we raided." He told his American brother that headhunting had never ceased. Almost all the males of the village had taken heads recently and were proud of this; Rosaldo was shocked. A few months later, he was drafted for the Vietnam War but he refused to go. He expected that his Ilingot friends disapproved of his decision but he was again surprised:

> My companions immediately told me not to fight in Vietnam, and they offered to conceal me in their homes. Though it corresponded to my sentiments, their offer could not have surprised me more. Unthinkingly, I had supposed that headhunters would see my reluctance to serve in the armed forces as a form of cowardice. Instead, they told me that soldiers are men who sell their bodies. Pointedly they interrogated me, 'How can a man do as soldiers do and command his brothers to move into the line of fire?' (Rosaldo 1989, 63)

Rosaldo then realized that a central Western institution, modern war, was met by his Ilingot friends with the same moral disgust that headhunting had aroused in himself:

> The act of ordering one's own men (one's 'brothers') to risk their lives was utterly beyond their moral comprehension. That their telling question ignored state authority and hierarchical chains of command mattered little. My own cultural world suddenly appeared grotesque. Their earnest incomprehension significantly narrowed the moral chasm between us, for their ethnographic observation about modern war was both

aggressive and caring. They condemned my society's soldiering at the same time that they urged me not to sell my body. (Rosaldo 1989, 64)

The Ilingot reaction had changed dramatically Rosaldo's perception of his own cultural position and in so doing Rosaldo's experience showed acknowledgement of moral diversity together with persistent tension to universal values.

The intercultural approach considers people to be autonomous and responsible moral agents with cultures that are unique configurations of specific histories and exchanges. An ethnographically oriented methodology is needed to study the activities of individuals and social groups in their specific cultural contexts. Current research in cultural psychology (Cole 1995, 1996; Shweder 1991, 2003), linguistic anthropology (Duranti 1994, 1997; Ochs and Capps 2001), situated action (Clancey 1997; Hutchins 1995) and discourse and intercultural mediation (Mantovani 2008a, 2008b) has provided examples of the work that could be done to consider people in their everyday activities: making decisions, negotiating the meanings of situations and telling stories according to their interests and goals. Studying the many occasions in which the new media support and even create interchange within and across cultural borders is different if the researcher is aware of the framework he/she is using to understand the various concepts of cultures, differences, communities and identities enacted in the communication space created by the new media.

NOTES

1. In this study no evidence was found of an expected acuity of eyesight among natives to balance the assumed superior intellectual acumen of Europeans, but the hypothesis explored in the research nevertheless revealed its ethnocentric Western biases.

REFERENCES

Appadurai, Arjun. 1998. *Modernity at large: Cultural dimensions of globalization*. Minneapolis: University of Minnesota Press.

Bakhtin, Michail M. 1981. *The dialogic imagination*. Austin: The University of Texas Press.

Baumann, Gerd. 1996. *Contesting culture: Discourses of identity in multi-ethnic London*. Cambridge: Cambridge University Press.

———. 1999. *The multicultural riddle: Rethinking national, ethnic, and religious identities*. New York: Routledge.

Benhabib, Selya. 2002. *The claims of culture: Equality and diversity in the global era*. Princeton, NJ: Princeton University Press.

Berry, John W. 1997. Immigration, acculturation and adaptation. *Applied Psychology* 46:5–68.

———. 2001. A psychology of immigration. *Journal of Social Issues* 57:615–631.

Castells, Manuel. 1996. *The rise of the network society.* London: Blackwell.

Clancey, William J. 1997. *Situated cognition.* Cambridge: Cambridge University Press.

Clark, Andy. 1997. *Being there.* Cambridge, MA: MIT Press.

Clifford, James. 1997. *Routes: Travel and translation in the late twentieth century.* Cambridge, MA: Harvard University Press.

Cole, Michael. 1995. Culture and cognitive development: From cross-cultural research to creating systems of cultural mediation. *Culture and Psychology* 1:25–54.

———. 1996. *Cultural psychology. A once and future discipline.* Cambridge, MA: Harvard University Press.

Cole, Michael, John Gay, and Joseph A. Glick. 1971. *The cultural contexts of learning and thinking.* Cambridge, MA: Harvard University Press.

Dershowitz, Alan. 2004. *Rights from wrongs: A secular theory of the origin of rights.* New York: Basic Books.

Despret, Vinciane. 2001. *Ces emotions qui nous fabriquent: Ethnopsychologie de l' authenticité.* Paris: Les Empecheurs de Penser en Rond/Seuil.

Duranti, Alessandro. 1994. *From grammar to politics.* Berkeley: University of California Press.

———. 1997. *Linguistic anthropology.* Cambridge: Cambridge University Press.

Duranti, Alessandro and Charles Goodwin. 1992. *Rethinking context: Language as an interactive phenomenon.* Cambridge: Cambridge University Press.

Fortunati, Leopoldina. 2005. The mobile phone between local and global. In *A sense of place: The global and the local in mobile communication,* ed. Kristof Nyiri, 61–70. Wien: Passagen Verlag.

Geertz, Clifford. 1994. The uses of diversity. In *Assessing cultural anthropology,* ed. Robert Borofsky, 556–559. New York: McGraw Hill.

Hannerz, Ulf. 1992. *Cultural complexity: Studies in the social organization of meaning.* New York: Columbia University Press.

———. 2001. Thinking about culture in a global ecumene. In *Culture in the communication age,* ed. James Lull, 54–71. London: Routledge.

Hutchins, Edwin. 1995. *Cognition in the wild.* Cambridge, MA: MIT Press.

———. 1997. How a cockpit remembers its speed. *Cognitive Science* 19:165–188.

Jahoda, Gustav. 1992. *Crossroads between culture and mind.* New York: Harvester.

Lave, Jean. 1988. *Cognition in practice.* Cambridge: Cambridge University Press.

Likert, Rensis. 1932. The method of constructing an attitude scale. *Archives of Psychology* 140:44–53.

Mantovani, Giuseppe. 1996a. *New communication environments: From everyday to virtual.* London: Taylor and Francis.

———. 1996b. Social context in HCI: A new framework for mental models, cooperation and communication. *Cognitive Science* 20:237–269.

———. 2000. *Exploring borders: Understanding culture and psychology.* London and New York: Routledge.

———. 2002. Internet haze: Why new artifacts can enhance ambiguity in situations. *Culture and Psychology* 8:56–78.

———. 2004a. *Intercultura: E' possibile evitare le guerre culturali?* Bologna: Il Mulino.

———. 2004b. Defending cultural pluralism against imperial visions. *Contemporary Psychology* 49:756–759.

———. 2008a. *Discorso e contesto sociale: Metodi qualitativi per un mondo plurale.* Bologna: Il Mulino.

———. 2008b. Intercultura: La differenza nel cortile di casa. In *Intercultura e mediazione: Modelli ed esperienze per la ricerca, la formazione e la pratica,* ed. Giuseppe Mantovani, 7–25. Rome: Carocci.

Markus, Hazel R., and Shinobu Kitayama. 1991. Culture and the self: Implications for cognition, motivation and emotion. *Psychological Review* 98:224–253.

Ochs, Elinor, and Lisa Capps. 2001. *Living narrative: Creating lives in everyday storytelling.* Cambridge, MA: Harvard University Press.

Ochs, Elinor, and Laura Sterponi. 2003. Analisi delle narrazioni. In *Metodi qualitativi in psicologia,* ed. Giuseppe Mantovani and Anna Spagnoli, 131–158. Bologna: Il Mulino.

Rogoff, Barbara. 2003. *The cultural nature of human development.* New York: Oxford University Press.

Rosaldo, Renato. 1989. *Culture and truth: The remaking of social analysis.* Boston: Beacon Press.

Schwarz, Norbert. 1999. Self reports: How the questions shape the answers. *American Psychologist* 54:93–105.

Sen, Amartya. 2003. Why democracy is not the same as Westernization: Democracy and its global roots. *The New Republic* 229:28–36.

———. 2004. What's the point of democracy? *Bulletin of the American Academy of Arts and Sciences* 42(Spring):8–11.

Shweder, Richard A. 1991. *Thinking through cultures: Expeditions in cultural psychology.* Cambridge, MA: Cambridge University Press.

———. 2003. *Why do men barbecue? Recipes for cultural psychology.* Cambridge, MA: Harvard University Press.

Suchman, Lucy. 1987. *Plans and situated actions.* Cambridge: Cambridge University Press.

Triandis, Harry C. 1989. The self and social behavior in different cultural contexts. *Psychological Review* 96:506–520.

Vgotsky, Levs. 1978. *Mind in Society: Development of Higher Psychological Processes.* Boston: Harvard University Press.

2   From English to New Englishes

Language Migration Towards New Paradigms[1]

Maria Bortoluzzi

INTRODUCTION

In this chapter I focus on one of the languages that has migrated most in the last few centuries and, as a consequence, undergone such metamorphoses that it is now called by its plural name to represent its multifarious identities: from English to Englishes. The chapter will focus in particular on the new Englishes of migrations influenced and shaped by the use of ICTs. It will examine their global/local audiences and the phenomenon of virtual language migrants who use English varieties as linguae francae for specific purposes within communities who do not need to be physically in the same place but share virtual spaces of communication. Language is one of the main means, tools and products of human migration. At the interface between nature and technology, language is the result of human evolution and has always been a cultural instrument contributing to and representing development and change. It is the first and most complex communication technology humankind has developed, inextricably wired in the human brain and constitutive of it. Languages are also among the most relevant communicative technologies involved directly and indirectly in migrations; native languages that migrants bring to their new country become insufficient for their communicative needs whilst remaining one of the loci of psychological, social and cultural identity. The new language(s) migrants encounter are at first a communication barrier comprising an incomprehensible system of oral and written signs which then impact psychologically on the personality of individuals and groups; indeed, their linguistic inability to express themselves causes adults to revert to child-like verbal skills. The new language, however, is also the means to gain access to the new culture, the tool to interpret it and change it from within. Thus the result of migrations are bilingual, trilingual, multilingual groups of people who code-switch from one language or language variety to another, adapting to contexts and events but also reinterpreting and shaping them and in so doing influencing and changing the local communities. Languages are so engrained in our nature that we speak of them in human terms; they belong to families and form family trees with parent, daughter and sister languages. For instance Latin is the parent language to French and Italian which are referred to as daughters to Latin and sisters to each other; continuing the human metaphor, they can be mother tongues or second languages to their speakers. They evolve, become endangered and die; they migrate along with the individuals who use them and give birth to new linguistic varieties, groups and families. The dominant metaphor, which blends and confuses nature and culture, represents languages as living beings mirroring and shaping our life (Cameron 2007).

This chapter aims to show how recent research developments in the study of the spread of English varieties have contributed to challenging some long-established linguistic paradigms and provided new critical stances in the relation between language varieties and power, socio-cultural identities and creativity of communities of practice in the local and global contexts through ICTs.

The theoretical frameworks referred to in this study belong to recent developments in sociolinguistics (see, for instance, Milroy and Milroy 1999; McArthur 2002; Crystal 2000, 2003a, 2003b; Graddol et al. 2006; Coupland and Jaworski 2009); critical discourse studies and critical language awareness (Fairclough 2001, 2003, 2006; Blommaert 2005; van Dijk 2008; Machin and van Leeuwen 2007; Caldas-Coulthard and Iedema 2008; Tan and Rudby 2008) as well as studies of language in ICTs (Baldry and Thibault 2006; Johnson and Ensslin 2007a; Baron 2000, 2008; just to mention a few).

The chapter starts by outlining the socio-cultural development of English into Englishes as related to migrations and to ICTs, exploring how this has changed some fundamental and traditionally established research paradigms in language study. The second and third sections present crucial aspects of the migrations of English as a world language, an international language and a lingua franca in relation to issues of new cultural and social identities. The concluding section summarizes how the use of ICTs is shaping and is shaped by the continuous "migrations" of English and can challenge or enhance established paradigms of what is considered, at the present moment, the "working tongue" of the global village (Svartvik and Leech 2006, 1).

FROM ENGLISH TO ENGLISHES

The spread of English is a widely studied phenomenon and it represents a central issue in language investigation because it spearheaded the use of new frameworks of reference and research tools to come to terms with its complexity and diversity. According to Rudby and Saraceni (2006b, 5) English is "a truly global language" widespread across a variety of domains and world regions. However, as Phillipson and Skutnabb-Kangas (1999, 22) argue, "Languages do not 'spread', just as 'countries' do not talk to each other." It is users that influence processes of globalization and localization. Languages migrate because users migrate both physically along economic and geographical routes and/or virtually along the routes offered by ICT. Languages spread not because of their own intrinsic qualities (a misunderstanding which is rather entrenched) but because they are one of the most relevant tools of communication for communities who have the economic, political and socio-cultural power to promote their language variety and succeed in doing it. So the question is one of "agency" and "power" of groups of people who use a certain language variety and manage to impose it on other groups. In migration, languages are potential barriers to communication and also the first relevant points of negotiation between the migrant and the local communities: learning a new language and maintaining one's own native language becomes a challenge and a power struggle involving deep-seated values such as cultural and self-identity.

The relevance of languages is, among other aspects, related to quantitative aspects, which partly influence or reflect the power of negotiation of their users. The latest survey of *Ethnologue*, a database which lists and catalogues all the known languages in the world (Lewis 2009), counts 6,909 living languages; among them 473 are labeled as nearly extinct and sociolinguistic studies predict that if this trend continues, in the 21st century 90% of languages will become endangered or extinct (Nettle and Romaine 2000; Tsunoda 2004; Duchêne and Heller 2007). Linguistic situations are in continuous evolution, but it is possible to state that the socio-historical events that favored in recent times the spread of English (and other dominant languages in the world) have been factors contributing to the disappearance of other languages. This tragic loss of languages and cultures experienced by the world population has enormously accelerated in the last two centuries (Duchêne and Heller 2007) creating the phenomenon of "vanishing voices" (Nettle and Romaine 2000). Using or losing a language impinges upon the way in which communities and individuals shape their social and self-identities through narratives and linguistically mediate "common dreams, fulfilled and unfulfilled imaginings" (Kramsch 1998).

The history of English starts in distant times with migrations and invasions of people from the northern regions of the European mainland towards Great Britain and Ireland. Germanic populations from the fall of the Roman Empire onwards, then the Vikings and in the 11th century the Normans, migrated (or raided and invaded the islands) bringing with them their own languages that were influenced by the pre-existing language substrata. At the origins of what we call English there are languages which have interacted intensely with others in an exchange of sounds, lexicon, structures and usages (Celtic, Latin, Germanic, Nordic, Norman, French, etc.). From its onset English, like all other languages, was a constellation of varieties due to regional, social, cultural, gender differences (Milroy and Milroy 1999). The southeast area of England around London, which had cultural, economic and political supremacy, contributed to consolidating its linguistic variety of English into a dominant standard; this variety eventually spread north and west in Britain and Ireland (whereas other English varieties and other languages, such as Gaelic, struggled to survive).

Over the centuries, when Britain became first a commercial and later a colonial power, the language migrated with peoples, cultural choices, goods, political and administrative institutions and financial decisions, and became the language which could instantiate, enact and impose British power abroad. English was shaping and being shaped by the Industrial Revolution and its policies; its migration from what had become the center (London) to the periphery (North America, Australia, New Zealand, India, etc.) transformed it (in the 19th century) into an overwhelmingly powerful world language, the language of the British Empire (Milroy and Milroy 1999; Crystal 2003a). The local languages, especially those which did not have a written codification, or a codification different from writing, tended to become endangered and more easily extinct. At the same time some English varieties developed into prestigious standards and became national languages after independence from the United Kingdom: American, Canadian, Australian, New Zealand, Indian, Ghanaian, Nigerian, Singaporean English to name a few (see McArthur 1998; Graddol et al. 2006).

The technological revolution of the 20th century driven and supported by industrial economy, multinational companies and global market forces took its impetus from the United States and, supported by other industrialized countries with a Western type of economy, contributed to spreading English towards new economic and technological migrations.

In the 1980s, a series of studies by Braj Kachru, a linguist of Indian origin, marked one of the paradigm shifts in linguistics and applied linguistics. Kachru (1985, 1989, 1992) envisages three concentric language circles: the Inner Circle consists of Britain plus all the countries where English originally migrated to and then became the native and official language (to the detriment of local languages); in the Outer Circle, English migrated but remains a second language for the local population; the Expanding Circle includes all the people who use English as a foreign language. English is no longer studied as a series of varieties derived from a single original variety, traditionally considered culturally "prestigious" and a "model" for non-native learners (standard British English), but rather a constellation of Englishes, a "plural" language. Simplifying a complex situation, countries such as the U.S., Canada, Australia, Britain—belong to the Inner Circle; Tanzania, Nigeria, India to the Outer Circle; and the Expanding Circle consists of anywhere English is taught as a foreign language.

Kachru's work has had an impact both on research and applied areas but, as Canagarajah (2000) argues, circles "leak". First, language varieties are in a continual state of change and hybridization. Secondly, market forces and the globalized economy powerfully influence local and global languages by the day. Thirdly, the widespread use of ICTs makes it impossible to keep such clear-cut linguistic categories of English varieties (see Crystal 2003a; Baron 2008; Goggin and McLelland 2009a).

Ethnologue (Lewis 2009) highlights that the twelve most-used languages as native languages in the world account for 50% of world speakers whereas all the other 6,900 existing languages account for the other half (and many of these are endangered and almost extinct). Western colonial and post-colonial powers are not the only responsible agents of change in this respect because other hegemonic languages (Chinese Mandarin, Hindi, Urdu, Arabic, Russian, Bahasa Indonesia, etc.) have greatly contributed to accelerating the disappearance of some local languages (Tsunoda 2004; Duchêne and Heller 2007; Harrison 2008). English, on the other hand, is now widely spoken as a second language and as a foreign language by such a fast-growing number of people in the world that the native speakers have been outnumbered by non-native speakers and the total number of users of English in the world (native and non-native speakers) is approaching or has already outnumbered Mandarin native users (Graddol 2006, 60).

Put simply, a native language (or "mother tongue") is the language acquired from very early stages of child development in a natural way at home and then used daily in the social context. A second language is generally a language not spoken at home but acquired early on in life or learned as an adult and regularly used for some specific exchanges and activities in the social environment (in the work place, education system, for administrative purposes, etc.). A foreign language is generally learned at school. Whereas the categories are theoretically separate, in practice there are no clear-cut boundaries between "second language" and "foreign language" speakers, and even the "native/non-native" distinction becomes blurred in multilingual contexts such as India and Singapore, where some supposedly non-native speakers are familiar with English from an early age, and use the language routinely.

The fastest-growing section of English users can be found in the Expanding Circle, which obtained its recognition as a "linguistic entity" in the post-colonial age; it is porous like the other circles and, along with the Outer Circle, it contributed to tilting yet another linguistic balance and shifting another paradigm. According to Graddol, today the number of native speakers seems to be less relevant in providing a world language status, whereas the number of second and foreign speakers is increasing in importance (2006, 60). The crucial change for English has been the shift from widely accepted (albeit controversial and fragmented) native speaker models to seemingly counterintuitive non-native speaker models; but what does the non-native speaker model represent? The issue is particularly relevant for Englishes because, currently, the number of non-native users has by far outnumbered the native speakers by three to one (Crystal 2005, 507). Englishes have migrated far and wide with multinational organizations, business and service-based economy, banks, international publishing houses, media channels and so on. The consequences of this diaspora of Englishes from the native speaker "center(s)" to the non-native speaker "periphery" and back is a continuous movement of interests and ideas which has contributed to the present status of the most learned foreign/ second language in the world.

POSITIONING ENGLISHES WORLDWIDE

The present-day situation of English at the global level is not univocally interpreted by linguists. The spread of English is seen by some in positive terms because it gives access to a range of media, international business, scientific and other academic communication (Graddol et al. 2006, 14). Crystal refers to the "linguistic pluricentrism" of English in the 21st century with its multiple "ownership" and stakeholders (Crystal 2005, 507), and de Swaan (2001) sees the present-day linguistic situation in the world as a hierarchical model where English has become the "globally dominant language", the "hyper-central" language of the world.[2]

Some linguists (such as de Swaan 2001; Svartvik and Leech 2006) also see in a positive light the fact that Englishes are today a pluricentric linguistic entity which gives voice to a variety of communities (more or less central and powerful, more or less mobile and valued), allowing them to communicate across different social, linguistic, cultural and economic backgrounds. However, a far more critical point of view sees Englishes as representing and reinforcing the dominant aspects of global socio-cultural and economic power: Western affluent societies, industrialized neo-colonial powers, world neo-liberalism, service-based, consumerist society, the global empire of corporations and so on (Canagarajah 2000; Phillipson 2008a). The two positions are neither mutually exclusive nor by any means clear-cut and entirely separate, as will be seen in the following sections.

THE NON-NATIVE LANGUAGE MIGRATION
TOWARDS NEW IDENTITIES

The previous section dealt with the socio-cultural shift in the identity of English to Englishes and its users: English has been appropriated by the former British colonies to become post-colonial varieties or World Englishes (McArthur 2002), with their own standardized forms, vernaculars and rich, dynamic and prestigious literatures. Englishes have moved a step further, breaking the non-native barrier, taking up new, controversial and migrant identities of second/foreign language (from now on "second language" or L2) and called with different names according to the emphasis given to the issue, i.e. English as an International Language (EIL), English as Lingua Franca (ELF), Global English and so on. The forces at play are both centrifugal and centripetal: in the former, different migrant varieties of English are becoming "nativized" and are diverging towards mutual incomprehensibility thus limiting their role as a global language of communication, envisaging the possibility of Englishes changing into different languages (as happened to Latin and the Romance languages). The strong centripetal forces however, such as the international press, multinational corporations, socio-political bodies, world travel and tourism seem to reinforce the need for a mutually comprehensible form of English (Rudby and Saraceni, 2006a). Interestingly, ICTs are the tools that empower both divergence and convergence at the same time; they can contribute to the migration of English as a global language, but they also facilitate the use of varieties contributing to fragmentation and diversification within the linguistic diaspora.

Jenkins has used ICT tools to study the evolution of the "migrant" varieties of English in large corpora of what she calls English as an International Language (Jenkins 2003a, 2003b) and then English as Lingua Franca (2007). She remarks that linguists have traditionally regarded the changes and "deviant" language used by native speakers as a sign of creativity, whereas the linguistic deviations of non-native speakers have usually been considered erroneous. Jenkins wishes to see EIL become "the language of the *self* for its L2 speakers;" that is, the language of identification and not merely for transaction and information (Jenkins 2003b, 11–13). Mutual intelligibility is the key issue for a lingua franca that has migrated so extensively: Jenkins found both followers and detractors when investigating the features of the "Lingua Franca Core".

Other linguists who have widely explored and described English as Lingua Franca are Seidlhofer and her collaborators. Her computerized corpora of linguistic data from non-native users have been gathered to describe the main characteristics of ELF use in specific contexts and communicative events (both oral and written). The project (ICT applied to language analysis) is ongoing and can be accessed online, along with the research articles stemming from the joint efforts of Vienna and Oxford University (VOICE Web site; Seidlhofer 2004, 2009).

The implications of the research on ELF (the geographically and virtually migrant language *par excellence*) impinge upon both theoretical aspects dealing with the identity of a language and the shift of research paradigms in applied linguistics and education: if the model is no longer the native speaker but "migrant" varieties, who or what is the model for English language teaching and learning? The question entails a series of problems related on the one hand to the enormous economic interests of teaching, learning and using English around the world, and on the other setting common standards of achievement. The scholars working on ELF are not looking for rules to teach or a non-existent monolithic linguistic entity to describe, but they are trying to capture some recurrent linguistic features influenced by the communicative event, the purpose of the communication, the socio-linguistic background of the users and so on. So far research has only offered tentative answers to understanding more about how bilinguals communicate in context (Prodromou 2006). Indeed some linguists are rather critical towards the notion of English as Lingua Franca. Canagarajah (interviewed by Rudby and Saraceni, 2006a) states that he is suspicious of English as Lingua Franca as an entity because post-colonial communities have already been creatively negotiating the place of Englishes in their lives for quite a while and they are changing it to their preferred cultural and linguistic practices. Codemixing, bilingual communicative strategies like codeswitching and the use of local accents and idioms are manifestations of this process of nativization, which he defines as the "micropolitics" of post-colonial resistance.

Phillipson (2003) remarks that the label "lingua franca" obscures the fact that communication is asymmetrical when native speakers are involved in it: Esperanto is a real lingua franca for everybody, not English—but Esperanto has never caught on. The global socio-cultural and political migrations English has undergone in recent decades have transformed it into the "glocal" language of power, dominant economy, media, academia, culture and international warfare, rather than a seemingly neutral lingua franca (Phillipson 2008b, 250). Englishes are the result of massive migrations of different kinds and represent the coexistence of national communities of speakers, global communities of corporations and international organizations, heterogeneous "migrant" communities of ICT users. As Machin and van Leeuwen (2007, 2) remark, the global media industries have responded to this changeable situation by deliberately creating diversity, producing global media in "local" languages and integrating "local" content in a variety of verbal and non-verbal texts. To add a further layer of complexity, global media products are not interpreted everywhere in the same way, but they are resemiotized and experienced differently as "glocal" communication. Thanks to ICTs, the migration of Englishes (as well as other languages) has become even more fragmented and widespread, contributing to yet another paradigm shift.

ENGLISHES AND DIGITAL NATIVES (AND NON-NATIVES)

As Baron (2008, 226) writes, "Contemporary language technologies are poised to redefine our longstanding notions of what it means to communicate with another person." For instance places become virtual and the "sense of place" must be redefined. In a situation where users tend to be continuously online (by means of cell phones and computers), filtering out the opportunities for communicative exchanges and deciding when we are "available" to communicate becomes more and more relevant ("controlling the volume", Baron 2008). The links between languages and cultures are less easy to identify using traditionally established criteria (Kramsch 1998). The concept of "communities" cuts across traditional borders and categories such as national, cultural, geographical, linguistic, social, etc. (Loos, Mante-Meijer and Haddon 2008). ICTs have become so naturalized in present-day communication that language categories are used to define their role in society: users are called "digital natives" or "non-natives" depending on the age they acquired/learned digital skills. ICTs have also become the locus of digital migration whereby individuals and communities meet in the virtual worlds of social networks, form virtual communities across space and time and communicate in languages which are not native. Alternatively, ICT users communicate in their own language when they are far away from their home community, contacting home and other migrants. Englishes have been at the core of these continuous changes facilitated and shaped by ICTs: on the one hand as the dominant language of the world media and economic corporations, and on the other in the proliferation of communities (migrant and local) that appropriate English varieties as tools of communication.

Integral to these changes is the electronic textuality that transforms verbal communication by offering opportunities for textual and discoursal hybridity, "which are specialised for trans-national and interregional interaction" (Fairclough 2006, 3; see also Machin and van Leeuwen 2007; Johnson and Ensslin 2007a). Among the linguists who have explored the changes English is rapidly undergoing in the new media, Baron (2000, 2008) and Crystal (2003a, 2005) underline that some seemingly novel characteristics have existed for a very long time in trends of language change. The electronic media and the Internet in particular, however, have promoted language variety, facilitated changes and are having a powerful impact on Englishes. They are blending conventions of oral and written affordances, widely exploiting multimodal aspects of communication (still and moving images, sounds, graphic devices, etc.), hypertextuality, virtual environments for communicative exchanges and so on. There is "vast potential for representing personal and local identities" (Crystal 2005, 520–521) and all this contributes to the rapid emergence of varieties different from the dominant standards of English. The "newest New Englishes" often become identified by blend names representing their origins, including: Singlish (Singaporean English), Tex-Mex and Spanglish (English and Mexican Spanish), Taglish (English and Tagalog in the Philippines) (Crystal 2003b, 2005; McArthur 2002; Tan and Rudby 2008).

Virtual and non-virtual migrant communities, native or non-native speakers of English varieties contribute to linguistic evolution and standard and non-standard varieties of Englishes change and are changed by other languages. Different affordances are exploited in ICTs so that written and oral discourses and genres include and are shaped by them. These include visual, sound, kinetic and graphic contribution to communication, and geographical, cultural and ethnic boundaries that can be easily bridged on the virtual spaces of ICTs. In particular, the participative tools of Internet and mobile communication allow e-users to remap their identities into "e-communities", which cut across, reshape or (at times) maintain traditional boundaries, developing their own "idiolect". Within e-communities there is another traditional boundary which is often bridged or removed: text producers and text receivers are conflated into e-actors who collectively contribute to maintaining, promoting, challenging and resemiotizing multimodal discourses (Loos, Mante-Meijer and Haddon 2008; Fortunati et al. 2010).

Thus whereas the spread of ICTs has been taking place within "new capitalism" and its discourses (see Fairclough 2003, 2006), it has also been promoting a dynamic set of ideological frameworks which are not restricted to dominant discourses but also enable marginal agencies to surface, and potentially alter, previous hierarchical relations (Johnson and Ensslin 2007b, 13). Appropriation of conventions (both technological and discoursal) in an original and creative way on the part of the world sociocultural "periphery" contribute to interesting media usages and phenomena, highlighting the key dimension of human agency and responsibility.

The examples are too diversified to do justice to them here so I will mention only Web sites and social networks where users blend local languages (in some cases non-written or non-standard varieties) and varieties of English where post-colonial and/or formerly highly marginalized voices express themselves on global media. For instance, communities of Ghanaian storytellers and "hiplife" e-actors (Altin 2008) use the Web to make their voices literally heard and their performances are seen in blogs, podcasts and video-recordings uploaded on YouTube. This allows them to communicate with local and distant e-users to be read, seen and listened to by a wide audience and link up their new multimodal discourses with long-standing local traditions of oral storytelling and performances ("musicov-erbal genres") (Agawu 2003).

The question of agency and resistance to inequality lies at the core of another paradigm challenge which has been happening globally on the Internet during the last decade. The shift from the predominance of English to a more equitable share of Web sites in other languages is already happening on the Internet. Graddol (2006: 44–45) reports studies that show how in 2000 the number of Web sites in English was 51.3% with all other languages sharing the remainder, whereas in 2005 the percentage of Web sites in English had dropped to 32%. Statistics from 2007 report that the number of English-language users is now a minority: 29.5% with a falling trend, whereas China has now the largest number of Internet users, after having been for a while second only to the United States (Goggin and McLelland 2009b, 4).

Leaving aside the well-established socio-political and economic advantages English has accumulated in its long history as a global language, there are also advantages at the technological level. An enormous quantity of information is already stored in the language and, even more remarkably, its apparently unequal power lies in the architecture of the Internet because

URLs and domain names are in Roman script, which search engines mainly use to retrieve Web sites. Whereas research remains still "Western-centric", "There has been a growing sense of the multilingual nature of the Internet," (Goggin and McLelland 2009b, 5) and work such as Goggin and McLelland's (2009a) tries to offer insights into the local realities of the global network and "reverse the flow of influence from the 'margins' to the 'center'" (Goggin and McLelland 2009b, 14).

CONCLUDING REMARKS

This chapter has outlined some of the most relevant and recent linguistic "migrations" that users of English (natives and non-natives) have promoted, co-constructed or undergone. In the last decades the role of ICTs in language migrations and in the co-construction of new Englishes has been crucial in accelerating and transforming processes of language change that would have been much slower or impossible in the past. Some conventionally-established paradigms have shifted, offering new perspectives for language research especially in the field of ICTs. English studies include today a constellation of Englishes and New Englishes. English varieties reflect an ever-changing kaleidoscope of cultural, ethnic and community (virtual and non-virtual) usages. Native-user usages are complemented and influenced by non-native-user usages. English's dominant position as the language of the Internet has not gone unchallenged and globalizing centripetal forces in language issues contrast with and complement local centrifugal forces as less dominant languages become more and more represented on the Internet. ICTs have contributed to transforming e-users into virtual migrants forming communities across space and traditional boundaries; at the same time they have helped non-virtual migrants to maintain links with their community of origin and establish new links with the wider communities of Web users in "the endless search for belonging to the constantly changing other" (Krzyżanowski and Wodak 2008, 97).

None of these issues is without controversy and each one carries with it the need for further research and new research tenets, including the acceptance, on the part of the academic community, that plurilingualism should be seen as an asset rather than a problem. Academic research is one of the areas in which English monolingualism is most dominant and (apart from a few exceptions) mostly unchallenged (Ammon 2007). This is yet another convention that the world of academia needs to renegotiate towards a more equitable communication where there is place, respect and prestige for scientific research in what are now "subaltern" languages. Respect for diversity and human rights (for all the migrant, local and virtually migrant communities) must include respect for linguistic diversity and linguistic rights.

NOTES

1. I am very grateful to Keith Mitchell for devoting time to reading my work and giving me precious feedback.
2. On the second tier of the hierarchy there are a few "super-central" languages including French, Spanish, Russian, Chinese, Japanese, Arabic, Hindi, German and Portuguese (many belong to former colonial or regional empires). The great majority of languages (98%) belong to small ethnic groups and have little or no political relevance.

REFERENCES

Agawu, Kofi. 2003. *Representing African music: Postcolonial notes, queries, positions.* London: Routledge.

Altin, Roberta. 2008. Mediascape africani: Dai "talking drums" all' "hiplife" ghanese. *Aut Aut* 339:179–194.

Ammon, Urlich. 2007. Global scientific communication: Open question and policy suggestions. In *Linguistic inequality in scientific communication today: AILA Review*, ed. Augusto Carli and Urlich Ammon, 20:123–133.

Baldry, Anthony, and Paul Thibault. 2006. *Multimodal transcription and text analysis.* London: Equinox.

Baron, Naomi Susan. 2000. *Alphabet to email.* London: Routledge.

———. 2008. *Always on: Language in an online and mobile world.* Oxford: Oxford University Press.

Blommaert, Jan. 2005. *Discourse.* Cambridge: Cambridge University Press.

Caldas-Coulthard, Carmen Rosa, and Rick Iedema, eds. 2008. *Identity trouble: Critical discourse and contested identities.* Basingstoke, Hampshire: Palgrave Macmillan.

Cameron, Deborah. 2007. Language endangerment and verbal hygiene: History, morality and politics. In *Discourses of endangerment*, ed. Alexandre Duchêne and Monica Heller, 268–285. London: Continuum.

Canagarajah, Suresh. 2000. *Resisting linguistic imperialism.* Oxford: Oxford University Press.

Coupland, Nikolas, and Adam Jaworski, eds. 2009. *The new sociolinguistic reader.* Basingstoke, Hampshire: Palgrave Macmillan.

Crystal, David. 2000. *Language death.* Cambridge: Cambridge University Press.

———. 2003a. *English as a global language.* 2nd ed. Cambridge: Cambridge University Press.

———. 2003b. *The Cambridge encyclopedia of the English language.* 2nd ed. Cambridge: Cambridge University Press.

———. 2005. *The stories of English.* 2nd ed. London: Penguin.

de Swaan, Abram. 2001. *Words of the world: The global language system.* Cambridge: Polity Press.

Duchêne, Alexandre, and Monica Heller, eds. 2007. *Discourse of endangerment.* London: Continuum.

Fairclough, Norman. 2001. *Language and power.* 2nd ed. London: Pearson Education.

———. 2003. *Analysing discourse: Textual analysis for social research.* London: Routledge.

———. 2006. *Language and globalization.* London: Routledge.

Fortunati, Leopoldina, Jane Vincent, Julian Gebhardt, Andraž Petrovčič, and Olga Vershinskaya, eds. 2010. *Interacting with broadband society.* Berlin: Peter Lang.

Goggin, Gerard, and Mark McLelland, eds. 2009a. *Internationalizing Internet studies.* New York: Routledge.

———. 2009b. Internationalizing Internet studies: Beyond anglophones paradigms. In *Internationalizing Internet studies,* ed. Gerard Goggin and Mark McLelland, 3–17. New York: Routledge.

Graddol, David. 2006. *English next.* The British Council. http://www.britishcouncil.org/learning-research-english-next.pdf (accessed October 30, 2009).

Graddol, David, Dick Leith, Joan Swann, Martin Rhys, and Julia Gillen, eds. 2006. *Changing English.* London: Routledge.

Harrison, David. 2008. *When languages die: The extinction of the world's languages and the erosion of human language.* Oxford: Oxford University Press.

Jenkins, Jennifer. 2003a. *World Englishes.* London: Routledge.

———. 2003b. Respecting diversity, promoting intelligibility: The challenge for English as an International Language. *Perspectives, A Journal of TESOL-Italy* 30(2):11–21.

———. 2007. *English as Lingua Franca: Attitude and identity.* Oxford: Oxford University Press.

Johnson, Sally, and Astrid Ensslin, eds. 2007a. *Language in the media: Representations, identities, ideologies.* London: Continuum.

———. 2007b. Language in the media: Theory and practice. In *Language in the media: Representations, identities, ideologies,* ed. Sally Johnson and Astrid Ensslin, 3–22. London: Continuum.

Kachru, Braj. 1985. Standards, codification and sociolinguistic realism. In *English in the world: Teaching and learning the language and its literatures,* ed. Randolph Quirk and Henry Widdowson, 11–30. Cambridge: Cambridge University Press.

———. 1989. Teaching world Englishes. *Indian Journal of Applied Linguistics* 15(1):85–95.

———. 1992. *The other tongue: English across cultures.* 2nd ed. Urbana, IL: University of Illinois Press.

Kramsch, Clare. 1998. *Language and culture.* Oxford: Oxford University Press.

Krzyżanowski, Michal, and Ruth Wodak. 2008. Multiple identities, migration and belonging: "Voices of migrants". In *Identity trouble: Critical discourse and contested identities,* ed. Carmen Rosa Caldas-Coulthard and Rick Iedema, 95–119. London: Palgrave Macmillan.

Lewis, Paul, ed. 2009. *Ethnologue: Languages of the world.* 16th ed. Dallas, TX: SIL International. Online version: http://www.ethnologue.com/ (accessed: October 30, 2009).

Loos, Eugène, Enid Mante-Meijer, and Leslie Haddon, eds. 2008. *The social dynamics of information and communication technology.* Aldershot, Hampshire: Ashgate.

Machin, David, and Theo van Leeuwen. 2007. *Global media discourse.* London: Routledge.

McArthur, Tom. 1998. *The English languages.* Cambridge: Cambridge University Press.

———. 2002. *Oxford guide to world English.* Oxford: Oxford University Press.

Milroy, James, and Lesley Milroy. 1999. *Authority in language.* 3rd ed. London: Routledge.

Nettle, Daniel, and Suzanne Romaine. 2000. *Vanishing voices: The extinction of the world's languages.* Oxford: Oxford University Press.

Phillipson, Robert. 2003. *English-only Europe? Challenging language policy.* London: Routledge.

———. 2008a. The linguistic imperialism of neoliberal empire. *Critical Inquiry in Language Studies* 5(1):1–43.

———. 2008b. *Lingua franca* or *lingua frankensteinia?* English in European integration and globalization. *World Englishes* 27(2):250–284.

Phillipson, Robert, and Tove Skutnabb-Kangas. 1999. Englishisation: One dimension of globalisation. *English in a changing world [L'anglais dans un monde changeant]: The AILA Review* 13:19–36.

Prodromou, Luke. 2006. Defining the "successful bilingual speaker" of English. In *English in the world: Global rules, global role,* ed. Rani Rudby and Mario Saraceni, 51–70. London: Continuum.

Rudby, Rani, and Mario Saraceni, eds. 2006a. *English in the world: Global rules, global roles.* London: Continuum.

———. 2006b. Introduction. In *English in the world: Global rules, global roles,* ed. Rani Rudby and Mario Saraceni, 5–16. London: Continuum.

Seidlhofer, Barbara. 2004. Research perspectives on teaching English as a lingua franca. *Annual Review of Applied Linguistics* 24:209–239.

———. 2009. Common ground and different realities: World Englishes and English as a lingua franca. *World Englishes* 28(2):236–245.

Svartvik, Jan, and Geoffrey Leech. 2006. *English, one tongue, many voices.* Basingstoke, Hampshire: Palgrave Macmillan.

Tan, Peter, and Tani Rudby, eds. 2008. *Language as commodity.* London: Continuum.

Tsunoda, Tasaku. 2004. *Language endangerment and language revitalization.* The Hague: Mouton de Gruyter.

van Dijk, Teun A. 2008. *Discourse and power.* Basingstoke, Hampshire: Palgrave Macmillan.VOICE (Vienna-Oxford International Corpus of English). 2009. http://www.univie.ac.at/voice (accessed October 30, 2009).

3 Frame Setting of Contestable Categories

The Construction of Multiracial Identity in the Mass Media

Alice Robbin

INTRODUCTION

The history of the United States is one of nearly continuous migration, beginning first with slave ships that brought Africans to work on southern plantations, followed by large-scale immigration during the1800s through the first two decades of the 20th century and a massive second wave of immigrants after changes to immigration and naturalization and civil rights laws during the 1960s. A demographic shift of momentous proportions has taken place, propelled largely by migration from developing countries, that has transformed the nation. The political and social consequences of these multiple waves of migration can be seen in the election of President Barack Obama, son of a white mother and Kenyan father, and in a newspaper report of a genealogical investigation of the slave origins of his wife Michelle Robinson Obama (Swarns and Kantor 2009).

It is the administrative record-keeping systems of the United States, in place since the first census of 1790, that provide us with a detailed profile of the composition of the American society. The 2000 decennial population statistics on race and ethnic origin reveal a demographic portrait of a diverse multiracial, multiethnic and multicultural nation and a census-to-census population increase that is the largest in the history of the United States (U.S. Bureau of the Census 2001) and significantly different in national-origin, racial and ethnic groups from earlier migrations (Quian and Lichter 2001).

Recent census estimates describe a foreign-born population of immigrants and naturalized citizens that is about 14% of the nearly 308 million people and has migrated from the five continents and represents all races and ethnic groups (U.S. Bureau of the Census 2009). Latin America accounts for 53%, Asia more than 27% and Europe 13% for the largest regions of originating countries. Of this foreign-born population, estimates for racial distributions indicate that whites account for nearly 50% and Asians almost 24%. Nationally, the number of native- and foreign-born people who have self-identified as "white" has seen a decline as a result of the in-migration of persons of other races. Forty-seven percent of this for-eign-born population defines its ethnic origin as Hispanic or Latino (Grieco and Cassidy 2001). The number of persons who self-identify as "more than one race" has seen the largest increase in the history of recording popula-tion statistics, estimated at 2.4%, or 6.8 million people according to the 2000 Census (Jones and Smith 2001).

Racial and ethnic group intermarriage is deemed the final stage of assim-ilation. Although overall, racial intermarriage increased from about 1% in 1970 to 6% in 2000 (Simmons and O'Connell 2003), newly arrived immigrant communities have tended to marry within and across pan-eth-nic native and immigrant groups (Quian and Lichter 2001). Intermarriage is less likely to occur with whites, in contrast to "the early waves of Euro-pean immigrants" whose "high rates of intermarriage with other White immigrants and natives tended to erode ethnic or national-origin group identities" that "helped foster a sense of 'Whiteness' as a distinctive ethnic or cultural group, and it hastened the process of Americanization" (Quian and Lichter 2001, 290, 291).

This second wave of immigration has significantly altered the racial and ethnic composition of many cities and suburbs and even entire states (Fong and Shibuya 2005). While reinforcing old patterns of assimilation, it has also created new patterns of immigrant settlement and incorporation and ethnic enclaves and economic "niches" (Logan and Alba 1999; Logan et al. 2000; Kasnitz, Mollenkopf and Waters 2002). Assimilation has become "segmented incorporation".

At the same time, however, because the 1960s incorporation has coexisted uneasily with the ideology of civil rights as the expression of political equality for all races and ethnic groups and with the philosophy of "diversity" as an expression of respect for persons. These compet-ing philosophies have contributed to resurrecting and (re)creating social conflict that has led to intense public policy debates in the late 20th and early 21st centuries that resemble the debates during the late 19th and early 20th centuries when state and federal governments applied strict immigration quotas, restrictions and exclusion. Social conflict is played out in the political arena of competing groups and policy networks and in the government agencies that are responsible for maintaining administrative record-keeping systems that record the characteristics of the population.

This migration leading to incorporation of racial and ethnic groups in the United States provides the historical context for understanding the political controversy about the reclassification of racial and ethnic data that took place during the 1990s and would subsequently influence how race became defined in the new millennium. A policy document known as "Standards for Maintaining, Collecting, and Presenting Federal Data on Race and Ethnicity" governs the United States government's prac-tices for civil rights compliance, general administrative record keeping and statistical data collection and reporting of race and ethnic data. This standard underwent a public review between 1993 and 1997 to determine whether the racial and ethnic group categories should be revised for the 2000 census and other government record-keeping systems. Beginning in the early 1990s, a national campaign to add a "multiracial" category created a symbolic "place at the table" for multiracial groups that was instrumental in framing the policy and research agenda of U.S. government agencies.[1] Multiracial and multiethnic groups skillfully utilized the mass media to mobilize adherents and gain sympathy for their position.

How the controversy, called a "thorny issue" by one *Washington Post* staff writer (Skrzycki 1994) and a "prickly issue" by one *New York Times* letter writer (Rheinhart 1996), over the reclassification of racial and ethnic data was reflected in the press between 1993 and 1997 is examined in this chapter. This was, as one newspaper staff writer put it, a time when "traditional racial lines are blurring and new ones are being drawn" (McLeod 1994) to "produce the All-American stew" as the editorial of the *Oregonian* (1994) commented and a "melting pot society" (Decker 1995).

This chapter examines the content of the messages as conveyed by the press, focusing on the symbols and potent stimuli employed by the advocates and opponents of a "multiracial" category that illuminates the contested policy arena of racial and ethnic classification and the more profound policy debates about the multiracial, multiethnic and multicultural society. The analysis relies on a corpus of work on language as a symbol system and the role of the mass media in influencing meaning and evoking the political "realities" that people experience.

Mass media institutions mediate and powerfully shape the information that reaches the public and directly and indirectly influence what people know about political events (Edelman 1988; Gulati, Just and Crigler 2004). Yet the "world of facts" that are reported by the media do not have a "determinable meaning"; rather, "news reporting continuously constructs and reconstructs . . . problems, crises, enemies, and leaders" (Edelman 1988, 1). The mass media are a major source for people's opinions on political issues, providing "cues about the probable future consequences of political actions, with information about the sources and authoritative support for policies, and with the groups with whom they identify" (Edelman 1988, 3). "What we 'know' about the nature of the social world depends upon how we frame and interpret the cues we receive about that world" (Edelman 1993, 231).

News stories are constructed messages that provide information and frame the unfolding of events for the reader. This framing process, according to Gamson and Modigliani (1987), may be thought of as a "symbolic contest" that is "waged with metaphors, catch-phrases, and other condensing symbols that suggest a core frame for making sense of relevant events" (137). The stories, whether the journalist reporting events, opinion pieces of syndicated columnists, editorials written by a newspaper's reporter or member of its editorial board or letters to the editor, represent a joint production between their author and reader that resonate with the cultural themes of a society. In these stories we encounter ideas about a policy issue as "packages" that "comprise arguments, information, symbols, metaphors, and images" (Gamson and Modigliani 1987, 143) that are "presumed to affect how people understand, interpret, and react to a problem or issue" (Tewsbury and Scheufele 2009, 19).

Official documents that record the history of the policy process that led to the reclassification of racial and ethnic data are widely available. It should be noted that press reporting differs little if at all from what can be found in official documents that record the process; indeed, reporters extracted statements from these official documents as part of their news stories. What differs is, however, the framing of the news reports and the salience of those official documents to stakeholders in the contested policy process.

This chapter analyzes how the political controversy over the reclassification of racial and ethnic data and the efforts of supporters and opponents of a "multiracial" category was described and evaluated by reporters, commentators, government officials, experts and letter writer citizens who responded to the press reports. In other words, how the press writ large constructed the explanation for the Office of Management and Budget (OMB) policy decision should also be considered a joint production of the news. This controversy is described in terms of the packages that framed the arguments, information, symbols, metaphors and images that were presented in daily newspaper and wire services reporting.

The analysis focuses primarily in the categorization of the problem made by the framing of the policy process and OMB's policy decision. Employing Gamson and Modigliani's (1987) structure for analyzing a public policy issue, four "packages" are identified to describe the debate about the addition of a "multiracial" category: remedial action, political power, one nation indivisible and numbers as objective reality. The LexisNexis Academic database was the source for the analysis dataset that consists of 381 news reports, editorials, news analyses and letters to the editor that appeared in 74 local, regional and national daily newspapers for the time period between January 1, 1993 and December 31, 1997.[2]

Concluding remarks return to the continuing importance and salience of racial identity in the United States and the objectivity of the press in reporting the controversy. This analysis of the reporting by the press of the government review can be read as a companion to earlier analyses of the history of racial and ethnic categories in the U.S. census, development of minority population interest groups and their representation in the national statistical system and the contentious public review from the point of view of the participants who were witnesses at congressional and administrative hearings (Robbin 1999, 2000a, 2000b, 2000c).

THE POLITICS OF RACIAL CLASSIFICATION:
FRAME SETTING OF CONTESTABLE CATEGORIES

We create our understandings of events and interpret our social world through the categories we construct. Chia (2000) writes that "it is through this process of differentiating, fixing, naming, labeling, classifying and relating—all intrinsic processes of discursive organization—that social reality is systematically constructed" (413). Racial and ethnic categories are symbolically potent for their consequences for identity and public policy. "In order to be fully countable and thus remembered by the state, a person needs first to fit into well-defined classification systems" (Bowker 2006, 30). However, classification systems provoke political controversy when the symbolic universe of language opens up to permit new conceptions of identity. The history of press reporting on the addition of a "multiracial" category is a case study in the politics of racial classification.

There is a consistency in the narratives that filled the pages of the newspapers and wire service reports and the themes that were articulated in all the documents contained in the analysis dataset. Newspaper reports, editorials, staff commentary, opinion pieces written by national syndicated columnists and responses by letter writers to news reports focused on images and definitions of self and America, the debate and controversy and the implications and consequences of policy change on citizens. Following Gamson and Modigliani (1987), the analysis examines the "package" and relies on the wording and on direct quotes to represent what was published in the press.[3]

1   Remedial Action and Identity

The core issue of this package is what the multiracial category represents in the body politic; that is, how to redress exclusion and invisibility in the society, which was the situation with census forms and other record-keeping systems. No longer needing to deny one part of oneself is a consistent story that is told by the advocates for a multiracial category. But it is one that was strongly opposed by minority populations. Yet there is an irony, noted by both advocates and opponents of a multiracial category, thus reinforcing our understanding that a symbol could be interpreted in more than one way, as Edelman (1988, 1993) pointed out.

In describing the upcoming Washington, D.C. Multiracial Solidarity March in Washington, D.C. in July 1996, one refrain dominated the discourse of multiracial groups. The march's organizer, Charles Byrd, son of a white father and a black mother, described the multiracial category as an "opportunity to proudly affirm a self-determined identity" (Moscoso 1996). Susan Graham of Project Race argued that the "census form was the correct place to make a personal, and sometimes political statement about one's racial makeup" because "the term was extremely important, especially to children" (Shepard 1997). And her son, quoted in the same newspaper article about his testimony at a congressional hearing two months earlier, said that "the 'other' on the 1990 census form 'makes me feel like a freak or a space alien'" (Shepard 1997).

Throughout the four-year period, the discourse was consistently about "visibility" and "invisibility", echoed by leaders of advocacy groups, parents of multiracial and multiethnic children and members of other minority population groups. "It's the children who don't have a voice," Ruth White, mother of six multiracial children, said. "The government is telling them they have to choose one race or the other. We say include everything, don't take away anything" (Flores 1995). Although referencing Middle Eastern immigrants, Helen Samhan of the Arab American Institute noted the sense that the current categories represented a "lack of recognition" and that "there's almost a sense of being ignored or invisible" (Sirica 1995). As Project Race advocate Graham was to put it in nearly every press report that quoted her, "It's a recognition of people's full heritage" (Foster 1995). Parents of multiracial children would tell reporters, in one way or another "We're trying to instill a sense of pride in who he is" because "they don't want him to assume skin color automatically conveys a person's background" (Mailander 1995).

Dickerson (1997), a member of the editorial board and staff writer for the *Atlanta (GA) Journal-Constitution*, concluded his editorial by pointing out that the multiracial category would "allow African Americans, officially for the first time, to embrace who they really ethnically are." It was above all, however, the sense of "feeling and becoming whole," no longer having to "straddle the racial divide" (Wu 1996). As such, "Minorities ought to unite," said Chris Zorich, a biracial and defensive tackle for the Chicago Bears football team (Wu 1996).The president of the National Association for the Advancement of Colored People (NAACP) conveyed the opposition of African Americans to a new category called multiracial (Shepard 1997). Before he gave reasons for opposing the category, he reaffirmed the need to "support the right of individual self-identification and support self-determination in defining one's racial makeup." At the same time, however, he questioned whether "the census form [was] the correct place to make a personal, and sometimes political, statement about one's racial makeup" (Shepard 1997). Yet, to argue that no remedial action was to be taken was to deny the fact was that "most African Americans—even those who consider themselves black—are multiracial," G. Reginald Daniel, a sociologist at the University of California at Santa Barbara, pointed out (Foster 1995). But this irony was even more profound, calling forth identification with the equally abhorrent Nazi ideology. Cox News Service reporter Smith (1996) wrote that "[Susan] Graham's [Project Race] proposal had stirred suspicions and attacks from some black people, who claimed it reeks of racism and perpetuates the false notion that racial purity exists." The reporter Smith then called on the authority of a civil rights law professor: "If you are seriously wondering whether some people are 'multiracial' or 'mixed,' then you probably think that people who are not 'mixed' are 'pure,'" which to this law professor "smacks of Nazi ideology."

2   Political Power

The core issue of this package is what the multiracial category represents in terms of political power. Race categorization was critical to both multiracial advocates and minority population groups but for different reasons. For the multiracial advocate, the multiracial category resonated with efforts to right the wrongs of a racist and segregated society that had subjugated African Americans for centuries and with the need to provide a label for their biracial and multiracial children. The African American and other minority population groups, including Hispanic/Latino, Asian and American Indian, opposed a multiracial category because a new category would reduce the benefits that had been accrued by existing categories. For the advocates of a multiracial category, opposition by minority population groups reinforced second-class citizenship. For the opponents of a multiracial category, the category represented a loss of political power.

The irony of the stance of the minority population groups was unmistakable because categories were necessary to affirm their identity (Robbin 2000b). Ramona Douglas, president of the Association of Multi-Ethnic Americans (AMEA), was among many advocates who found the opposition from African Americans and other ethnic minority groups "truly ironic" (Dickerson 1997). Racism was responsible. Douglas pointed out that "a century-ago, the one-drop rule" was used by minorities "to keep them down" and "now, this same one-drop rule is being used by minorities to keep power and rights at the expense of my community" (Dickerson 1997). Dickerson concluded his editorial by pointing out that African American opposition "could stick with massa's view of who's black and who's not," thus implying that opposition only reinforced the historically subjugated status of black Americans. Byrd, organizer of the Multiracial Solidarity March on Washington, D.C. in July 1996, categorized his opponents as those who had a "separatist' ideology" (Moscoso 1996). Byrd was quoted by another reporter: "African American political leadership opposed the multiracial initiative" because it "would diminish its accrued political power. The more people who identify as black, the more power those leaders amass; the less political power and the less support for programs such as what remains of affirmative action" (Stewart 1996). Confirmation of the accuracy of Byrd's statement came from the Washington, D.C. office legislative director of the National Association for the Advancement of Colored People (NAACP). This NAACP figure was quoted by the Associated Press as saying that "some people who choose the multiracial designation could inadvertently 'disconnect' from the larger black coalition they now identify with. 'Will a new multiracial congressional district be set up out of this?

No.' he asked and answered, rhetorically. 'Now we are together as a large group, and to start to micro-define won't help'" (Virginian Pilot (Norfolk) 1997). The consequences would be significant: loss in access to governmental funds and political representation if the multiracial category were added to the census item on race (Moscoso 1996).

3   One Nation Indivisible

The core issue of this third package concerns the history of race (and ethnicity) in the United States. Most newspaper staff writers explained the salience of the issue of a multiracial category as a "rancorous debate" (Shepard 1997) and "heated issue" (Moscoso 1996) between advocates who were themselves or parents of multiracial or biracial children and opponents who were defenders of the current categories because they saw their racial and ethnic groups as losers in access to governmental funds and political representation if the multiracial category were added to the census item on race. But the issue goes far deeper into the psyche of the American people, reflecting the history of a society both racist and a melting pot of nationalities.

Carl Rowan (1997), nationally syndicated columnist, pointed out, "Children of black-white marriages wonder why they always are counted as 'blacks' when they are mostly Caucasian. They know that there is still a stigma to being called black, a holdover from old statutes decreeing that anyone of even one-sixteenth African descent, or with 'one drop of Negro blood,' was a 'Negro' whose rights were to be curtailed." Rowan went on to describe the history of the abusive treatment of blacks by slave masters.

One editorial writer suggested that "our race-torn culture might feel whole again. We may survive if we support the multiracial category" (Wu 1996). Another editorial writer had written several years earlier that "the multiracial box, first of all, is a better description of the reality that increasingly is becoming America. To pretend it doesn't exist simply offends truth. Someday, and inevitably, we'll all be checking the multiracial box, and I think that would be a good thing" (Byrne 1993).Nonetheless, their views were not uniformly shared. An *Atlanta (GA) Journal-Constitution* editorial, written following the release of the July 1997 OMB report that had rejected the multiracial category, quoted the government's report that a "multiracial category would no doubt add to racial tension and further fragmentation of our population" and that "no new 'groups' were needed in this country" (Matthews 1997). George Will (1997), a nationally syndicated moderate-conservative-leaning columnist, added his opposition to any racial categorization, arguing that "today the government concocts 'race-conscious remedies' such as racial preferences for conditions it disapproves. This encourages Americans to aggregate into groups jockeying for social space. Perhaps it would be best to promote the desegregation of Americans by abolishing the existing five census categories, rather than adding a sixth." The multiracial category was also opposed by those who spoke for a white and conservative majority. Editorial writer Warner (1997) of the conservative newspaper the *Orange County (CA) Register* saw it "as high time the land of the free abandoned the bankrupt policy of taking skin color and ethnic background closely into account when it deals with citizens." Warner connected the census and racial classification with government "intrusiveness"; thus, one way to reduce the power of government *cum* racial and ethnic groups would be to eliminate all categories. Demonstrating the extent to which the history of race in the United States continued to be a salient topic, letter writers to the *Washington Post* who advocated a multiracial category noted that categories were needed because the society was "not yet colorblind" and that "race had been a national obsession for centuries" (*Washington Post* 1997). Letter writers who opposed the multiracial category contended that all racial categories ought to be eliminated because "racial data had always been flawed" due to the society's more than two centuries of interracial relationships.At the same time, however, a multiracial category reflected its origins in a society of many races and many ethnic groups, which some saw as isolated communities but a society that was no longer. The "face of America is changing," said many people; we are a "rich stew" (Glassman 1997). We are, one letter writer wrote, "rapidly moving toward a blended race, and our next census should reflect this change" (*San Jose (CA) Mercury* 1997). Embracing a multiracial category suggested a way to "help break down the polarizing myth of bifurcated white and black categories that continues to sow racism in American society" (*Record (Bergen, NJ)* 1997).

4   Numbers as Objective Reality

The core issue of this package is the reliance on government for objective, unbiased statistics and on government as the official authority for imposing order on identity. Government statistics are significant organizing vehicles for contemporary society (Stone 1997). They represent official action and the penetration of seeming objectivity into everyday life that codifies human behavior and formalizes the social world. The authority of government statistics was readily in evidence: statistics were used by advocates and opponents of the multiracial category to support as well as to oppose any sort of categorization of race and ethnicity. Although the public review reflected a history of contesting government action and the role of government, the production and quality of the statistics themselves appear to have been very nearly unquestioned by supporters and opponents alike of a multiracial category. Although counting meant winners and losers in political representation, segregation and anti-discrimination programs and a host of government programs, as was pointed out by minority population groups that opposed the addition of a multiracial category, these groups never questioned the veracity of the numbers.

It is important to recognize that this reliance on the authority of government statistics was not new to the United States (Porter 1995). As a "calculating people" the United States had, even before it officially became a republic, developed measurement systems to count its people; its 1790 census was just another manifestation of the spirit of the Enlightenment that was enthusiastically institutionalized (Cohen 1982). Statistics provided unassailable evidence of the progress the young nation was making. Thus, it is unsurprising that a large number of news reports, editorial commentary by newspaper staff or syndicated columnist and letter writers who responded to published reports about the OMB public review quoted government sources and cited statistics that described emerging demographic trends in interracial couples, the growing number of multiracial and multiethnic children and social changes that the society was undergoing.The authoritative role of the government was reinforced throughout the period of the public review by reports of statistics on the growth of interracial marriage and recent demographic trends in race and ethnicity that resulted from changes due to immigration and birth rates in the population. News reports reported on the research program conducted by the government agencies and the results of the research that suggested whether category labels would or would not affect population distributions. News reporters quoted government officials involved in the interagency decision-making process. The government research program also drew its authority by relying on members of the academy, who were quoted in the news reports. And government's authority was further reinforced by support from minority population interest groups who employed the statistics to oppose the addition of a multiracial category and any changes in the race categories that had been in force since 1977.News reporting rarely discussed challenges to the meaning of the categories or recognized the ambiguity of the meaning of the categories. The news reporter Younge (1996) was one of the few, commenting that "the core problem of a category was that nobody has a clear definition of what [the multiracial category] means." Congressman Sawyer too saw the problematic nature of categorization. News reporter Sirica (1995) quoted the congressman: "With so much social change, the fight is not just over a 'small box.' Rather, the census boxes define 'who we think we are and are becoming, and therein lies the compelling force behind an awful lot of peoples' desire for something other than the crisp, hard definitions that exist right now.'" Counting "'doesn't represent accuracy,' Sawyer said. 'It represents an illusion of precision.'" Overall, however, the reader of these newspaper accounts of the policy process would conclude that the numbers were taken for granted, they were "real" and they were objective accounts of reality.

CONCLUDING THOUGHTS

The context of migration to the United States is central for understanding how administrative record-keeping systems to record race and ethnic group identity have evolved. Racial identity and racial practices become embedded in political institutions and in governmental processes and practices and are represented (or not) in the official classification system for racial and ethnic group data. Its categories are part of the collective consciousness and have dominated the historical and contemporary language about conceptions of race and ethnicity in the United States. History, culture, the organization of social life, politics and technology have contributed to making the classification system for racial and ethnic data and its categories and their boundary conditions a contested policy domain and its numbers a source of dispute, as it concerns "legitimation [power] and explanation [about the past as well as the present]" (Olnick and Robbins 1998, 108).

News reporting presented "an on-going negotiation among key actors in [a public policy] process" (Gulati, Just and Crigler 2004, 237). The mass media comprised of news reporters, editors and commentators, and others including government officials, politicians, interest groups and experts, were represented in press reports and played active roles in this process. Voice was given to all; to this extent, then, the press was "objective" and the premise of a partisan bias reveals none. Still, one might conclude that the number of news articles written by commentators, editorials by members of the press, the prominent role that opponents and experts played in the process and the description of the final decision by the Office of Management and Budget suggests that, ultimately, what counted was the authority of government, experts and minority population groups.Race was a significant policy issue during the 1990s, and race and immigration continue to be inextricably bound in the new millennium. The public debate continues unabated.

NOTES

1. I place quotation marks around the name of the multiracial category to reinforce the fact that it is socially constructed. Although I do not do so for all the other racial and ethnic category group names referenced in this chapter, it should be understood as such. These quotation marks emphasize the ideological and political character of the terms. Precisely because such categories become taken for granted as real, the reader is reminded of their arbitrary and historically contingent character.
2. A complete description of the construction of the analysis dataset is available from the author.
3. To avoid confusing the reader, this section does not put double quotation marks around "multiracial."

REFERENCES

Bowker, Geoffrey. 2006. *Memory practices in the sciences.* Cambridge, MA: MIT Press.
Byrne, Dennis. 1993. Racial box requirement is bad form. *Chicago Sun-Times,* April 27.

Chia, R. 2000. Discourse analysis as organizational analysis. *Organization* 7:513–518.

Cohen, Patricia. 1982. *A calculating people: The spread of numeracy in early America.* Chicago: University of Chicago Press.

Decker, Jonathan P. 1995. Florida draws new race line. *Christian Science Monitor,* September 25, 3.

Dickerson, Jeff. 1997. Opponents misguided; The time has come for "mixed race" category. *The Atlanta (GA) Journal-Constitution,* May 20, 12A.

Edelman, Murray. 1988. *Constructing the political spectacle.* Chicago: University of Chicago Press.

———. 1993. Contestable categories and public opinion. *Political Communication* 10:231–242.

Flores, Laura. 1995. Multiracial groups seek identity; U.S. census urged to add category. *Times Picayune (New Orleans, LA),* December 10, A12.

Fong, Eric, and Kumiko Shibuya. 2005. Multiethnic cities in North America. *Annual Review of Sociology* 31:285–304.

Foster, Shawn. 1995. Falling through racial cracks race: People defy categorization. *Salt Lake (UT) Tribune,* November 27.

Gamson, William A., and Andrè Modigliani. 1987. The changing culture of affirmative action. *Research in Political Sociology* 3:137–177.

Glassman, James K. 1997. Multiracial (the colorblind choice). *San Jose (CA) Mercury News,* June 12, 10B.

Grieco, Elizabeth M., and Rachel C. Cassidy. 2001. Overview of race and Hispanic origin. Census 2000 Brief, C2KBR/01–1. Washington, DC: U.S. Bureau of the Census.

Gulati, Girish J., Marion R. Just, and Ann N. Crigler. 2004. News coverage of political campaigns. In *Handbook of political communication,* ed. Lynda Lee Kaid, 237–256. Mahway, NJ: Lawrence Erlbaum Associates, Publishers.

Jones, Richard A., and Amy S. Smith. 2001. Two or more races population: 2000. Census 2000 Brief, C2KBR/01–6. Washington, DC: U.S. Bureau of the Census.

Kasnitz, Phillip, John Mollenkopf, and Mary C. Waters. 2002. Becoming American/Becoming New Yorkers: Immigrant incorporation in a majority/minority city. *International Migration Review* 36(4):1020–1036.

Logan, John R., and Richard D. Alba. 1999. Minority niches and immigrant enclaves in New York and Los Angeles: Trends and impacts. In *Immigration and opportunity: Race, ethnicity, and employment in the United States,* ed. Frank Bean and Stephanie Bell-Rose, 172–193. New York: Russell Sage Foundation.

Logan, John R., Richard D. Alba, Michael Dill, and Min Zhou. 2000. Ethnic segmentation in the American metropolis: Increasing divergence in economic incorporation, 1980–1990. *International Migration Review* 34:98–132.

Mailander, J. 1995. Complex lineage defies labels for many in U.S.; Schools grapple with race issue. *Times-Picayune* (New Orleans, LA), June 11, p. A.20.

Matthews, Richard. 1997. No new interest group; "Many races" is better than "multiracial". *Atlanta (GA) Journal-Constitution,* July 10, Journal edition.

McLeod, Ramon G. 1994. U.S. racial categories criticized by minorities. *San Francisco Chronicle,* July 15, A9.

Moscoso, Eunice. 1996. Push for multiracial census category sparks demonstration. *Atlanta (GA) Journal-Constitution,* July 20, 10A.

Olnick, Jeffrey K., and Joyce Robbins. 1998. Social memory studies: From collective memory to the historical sociology of mnemonic practices. *Annual Review of Sociology* 24:105–40.

Oregonian. 1994. Editorial section. July 12, B08.

Porter, Theodore M. 1995. *Trust in numbers: The pursuit of objectivity in science and public life.* Princeton, NJ: Princeton University Press.

Quian, Zhenchao, and Daniel T. Lichter. 2001. Measuring marital assimilation: Intermarriage among natives and immigrants. *Social Science Research* 30:289–312.

Record (Bergen, NJ). 1997. Off the shelf/politics/history. February 20, Y02.

Rheinhardt, Marilyn. 1996. Multiracial people must no longer be invisible; Diminishing us all. *New York Times,* July 12, 26.

Robbin, Alice. 1999. The problematic status of statistics on race and ethnicity: An "imperfect representation of reality". *Journal of Government Information* 26:467–483.

———. 2000a. Administrative policy as symbol system: Political conflict and the social construction of identity. *Administration & Society* 32:398–431.

———. 2000b. The politics of representation in the national statistical system: Origins of minority population interest group participation. *Journal of Government Information* 27:431–453.

———. 2000c. Classifying racial and ethnic group data: The politics of negotiation and accommodation. *Journal of Government Information* 27:129–156.

Rowan, Carl. 1997. Census misses mark on race. *Chicago Sun-Times,* November 2, 37.

San Jose (CA) Mercury News. 1997. Back-seat ordinance cannot be justified by current statistics. June 13, 11B.

Shepard, Scott. 1997. Moving from "Other" to "Multiracial"; Melting pot: Many pushing to establish new race category in census. *The Atlanta (GA) Journal-Constitution,* July 5, 9A.

Simmons, Tavia, and Martin O'Connell. 2003. *Married-couple and unmarried-partner households: 2000.* Census 2000 Special Report, CENSR-5. Washington, DC: U.S. Bureau of the Census.

Sirica, Jack. 1995. Puzzling over racial categories. *The Times Union (Albany, NY),* January 30, C1.

Skrzycki, Cindy. 1994. At OMB, a thorny issue comes in the colors of a rainbow. *Washington Post,* June 17, F1.

Smith, Starita. 1996. Racially "mixed" seek category in census; Minority groups balk at push for new classification. *Times-Picayune (New Orleans, LA),* June 2, A24.

Stewart, Scott. 1996. March seeks new race category. *Chicago Sun-Times,* July 7, 7.

Stone, Deborah. 1997. *Policy paradox: The art of political decision making.* 2nd ed. New York: W. W. Norton and Company.

Swarns, Rachel L., and Jodie Kantor. 2009. In first lady's roots, a complex path from slavery. *New York Times,* October 7, A1.

Tewksbury, David, and Dietram A. Scheufele. 2009. News framing theory and research. In *Media effects: Advances in theory and research* (3rd ed.), ed. Jennings Bryant and Mary Beth Oliver, 17–33. New York: Routledge.

U.S. Bureau of the Census. 2001. Census 2000 shows America's diversity. Press release, CB01-CN.61. Washington, DC: U.S. Department of Commerce.

U.S. Bureau of the Census. 2009. Selected characteristics of the foreign-born population by period of entry into the United States. Dataset: 2009 American Community Survey one-year estimates. Washington, DC: Bureau of the Census.

Virginian Pilot (Norfolk). 1997. Census changes; An end to "Other". April 9, B10.

Warner, Gary A. 1997. Counting the people. *Orange County (CA) Register,* July 18.

Washington Post. 1997. Census: Backtalk. June 22, final edition, C05.

Will, George F. 1997. Melding in America. *Washington Post,* October 5, final edition.

Wu, Olivia. 1996. Solidarity march is right step. *Chicago Sun-Times,* July 13, 17.

Younge, Gary. 1996. Multiracial citizens divided on idea of separate census classification. *Washington Post,* July 19, A03.

Theme 2

Gender and Generation Intertwining with Migrations

4 Grandmothers, Girlfriends and Big Men
The Gendered Geographies of Jamaican Transnational Communication

Heather Horst

INTRODUCTION

For nine months of the year, Winston and his girlfriend Donna live as a transnational couple. Donna, a woman who was born on the east coast of the United States, first met Winston on a trip to visit a few friends in rural Jamaica. While the initial trip was only a few weeks, Winston and Donna stayed in touch until Winston managed to acquire a visa to the U.S. for a few months of farm work. Winston did not take to farm work and eventually decided that he was most content living a relatively thrifty, "natural" lifestyle in the hills of Jamaica. Donna, by contrast, had a small business that she had developed and was committed to maintaining. Rather than break off their romance due to distance, they decided to stay involved through Donna's annual stay (three months) in the winter months as well as weekly and, at times, daily phone calls. As Winston describes:

> Yes it [the mobile phone] made a lot of difference. It set your mind more at ease. Sometime she get worried, but now as soon as she want to talk she just dials. She want to call me night day or anytime. I am doing my own work so I can stop at anytime. Sometime she calls at night or in the afternoon. She calls before she goes work. (interview with author, 2004)

While the distance presents its challenges and the cost of calls may inhibit the sense of full-time intimate community we see in other contexts (Ito, Okabe and Matsuda 2005), for many couples the ability to maintain romance and a household across national boundaries is becoming less remarkable in the digital age.

This chapter explores transnational relationships with particular attention to the gendered dynamics of the communication and connection in the context of extensive transnational migration (Levitt and Glick Schiller 2003; Basch, Glick Schiller and Szanton-Blanc 1994). As feminist scholars have argued, the construction of gender relations and ideologies fundamentally structures communication and within the context of transnationalism, gender emerges as a consistent and important variable in two ways. In the first instance, nation-states often decide who can migrate and why, based on categorical and structurally-shaped definitions of gender, what Mahler and Pessar (2001) describe as the "gendered geographies of power", a concept that accounts for the spatial and social scales, social location and the types and degrees of agency expressed and exercised in transnational spaces (Pessar and Mahler 2003). Historically Jamaican men have been recruited to the United States for farm work on sugar cane plantations in the south and fruit farming in the Northeast. Women, by contrast, were admitted for professional occupations such as nursing and, in the more recent past, teaching and service sector jobs in the tourism industry. While laws signed for family reunification and the acquisition of professional and advanced degrees by a segment of women in Jamaica have shaped such dynamics, these structural trends continue to influence how individuals and their families cope with the importance of migration in Jamaican life.

Just as migration is a gendered process, so too are the technologies that facilitate communication. As Fortunati and Vincent (2009, 7) illustrate, the production, circulation and use of technologies represent a "site of power . . . over nature . . . human beings and . . . of control and dominion," particularly in Western contexts (see also Turkle 1984; Fortunati and Vincent 2009; Wacjman 1991, 2004). The ownership, control over, provisioning and use of objects like mobile phones reflect particular notions of power and control. For example in previous work on mobile phones in Jamaica (Horst and Miller 2006; Miller 2009), men often described the mobile phone as a masculine device, drawing parallels between mobile phones, cars and guns, all objects that enable domination and control over space and people. By contrast, discourses surrounding mobile phone usage often reproduce notions of women as emotional and expressive, tied to the role of the mobile phone in the care work of the family and efforts to restrict men's freedom and mobility. Drawing upon ongoing ethnographic research between 2004 and 2008 focused upon rural and urban Jamaicans' transnational engagements, I attend to the ways in which provisioning associated with expressions of love and commitment have been transformed with the widespread appropriation of the mobile phone in Jamaica. Moreover, I reveal how the mobile phone becomes a lens for understanding the differential power, relations and dependency among rural Jamaicans and their families and friends who live outside the country. While these new media and technologies have enabled communication, they are also spaces where power dynamics and particularly gender dynamics emerge, extended and reproduced, reflecting what Sarah Mahler (1998, 78) has described as the "bumpy and discontinuous" character of transnational social fields.

THE INFRASTRUCTURE OF TRANSNATIONAL MIGRATION: COLLECT CALLS, ICAS AND PHONE CARDS

Before turning in more detail to the gender dynamics of ICT use in transnational social fields, a brief sketch of the infrastructure of transnational communication and the telecommunications landscape in Jamaica is necessary, given that "we cannot understand how people communicate across national borders . . . until we see how they are situated vis-à-vis access to communication technologies" (Mahler and Pessar 2001, 4). Prior to the mid- to late 1990s, telephone lines were present throughout Jamaica via public phone boxes and local businesses in rural areas. In urban centers, including Kingston, Montego Bay and smaller parish capitals, landlines ("house phones") were common in upper and middle class homes. Many Jamaicans living in rural towns received their residential lines in the late 1990s when Cable and Wireless Jamaica Ltd replaced Telecommunications of Jamaica Ltd (in February 1998) and made a concerted effort to bring phone lines to the towns in Jamaica's rural interior, although those living in the most remote districts were unable to obtain a house phone and continued to rely on the tenuous service provided by phone boxes. If they were fortunate to possess social capital and connections, others could arrange to borrow the phone of a member of their extended family, co-worker or church brother or sister. International calls relied upon an International Call Authorization System (ICAS) code, which was entered into the phone prior to each overseas call. In lieu of an ICAS code, which was not commonly acquired by lower and middle class Jamaicans, individuals who wanted to call abroad during this time period were able to purchase a "World Talk" calling card from Cable and Wireless. Like ICAS, calls were considered expensive (over J$25 a minute, or around US$0.40 before 2002). Others preferred to place a collect call and until 2001, collect calls remained the most popular way to initiate contact with family and friends living abroad. Vertovec (2004, 220) notes that calls placed between the U.S. and Jamaica increased 127% between 1995 and 2001.

In the spring of 2001 the Irish telecommunications company Digicel entered the Jamaican telecommunications market (Boyett and Currie 2004; Horst and Miller 2006). The arrival of the mobile phone, and Digicel in particular, was considered a boon to communication within the Jamaican extended and transnational families. From the perspective of many Jamaicans with friends and family living outside of Jamaica, the direct access to international calls using the same credit placed on their phones for local and national calls represented a valuable asset. Possession of a mobile phone enabled Jamaicans to initiate and pay for communication with their friends and family abroad. Indeed, in 2004 placing a phone call overseas using a mobile phone cost the same as the J$17 call to another company's mobile phone in Jamaica, a typical call abroad considered to be the equivalent of J$200, or just over US$3.[1] In our analysis of the numbers and names saved in mobile phone books (Horst and Miller 2005), families and friends who lived outside Jamaica continued to be one of the most valued names saved within phones and people express great distress when a number was disconnected, changed without a forwarding number, or did not work. In the following sections, I begin by tracing the ways in which the mobile phone shapes gendered practices, relationships and households across space and time, with attention to the ways in which notions of masculinity and femininity emerge in the motivations and patterns of use underpinning transnational communication.

GENDERING TRANSNATIONAL FAMILY LIFE

As noted in the introduction, the history of migration among Jamaicans continues to be gendered, both in terms of the categorical definition of gender mapped by states who give preference for particular forms of labor and the expression of masculinity and femininity tied to the movement of migrants across national borders. For example, in the 1960s twice as many Jamaican women as men migrated for service and factory work to the United States (Foner 1983; Brown 1997; Thomas-Hope 1992). In exchange for economic support, women left their children with extended family members (especially grandmothers) in order to access jobs as full-time domestics caring for the children of white women entering the workforce in the 1970s (Foner 2005; Soto 1987; Vickerman 1999). The practices that emerged in the post-1965 migration to the United States built upon a long history of child fostering in Jamaica wherein female relatives, such as aunts and grandmothers, care for the children left behind. While fathers do play an economic and emotional role and, in some instances, take on roles as primary caregivers, they tend to rely heavily on their female relatives who practice more of the day-to-day care work in exchange for the economic support within these transnational care networks (Hochschild 2000; Goulbourne and Chamberlain 2001; Olwig 1999; Thomas-Hope 1988; Thompson and Bauer 2000).

Although increased access to travel facilitated return visits with greater frequency, new information and communication technologies like mobile phones began to play an important role in the care work of the family at a distance. Parents who lived in the Cayman Islands, U.S., Canada, England and elsewhere used the mobile phone to gain direct contact with their young children living with relatives in the rural regions of Jamaica, and their children (and children's caregivers) also initiated calls to update parents about their children's wellbeing and needs. Parents and children "read" the receipt of money in a birthday card, payment of school fees or shipping containers (barrels) of rice, clothing, small household appliances, school supplies and other forms of provisioning as an authentic demonstration of love and care, despite the local critique of "barrel children", a term coined by social workers in the 1980s to describe children who have been left with family in Jamaica and whose primary source of connection with their mothers and/or fathers revolves around the receipt of money and consumer goods. In the context of the Philippines wherein access to mobile phones among youth and across most sectors of Filipino society is prevalent, children are active participants in their transnational childhood.

Parreñas argues that children in Filipino transnational families "hunger for emotional bonds with absentee parents and wish for the intimacies of everyday interactions" (Parreñas 2005, 133; see also Olwig 1999). In contrast to recent studies of teenagers in the U.S. and Western Europe who seek autonomy and independence from their parents (Horst 2009; Ling and Yttri 2002), many of the teenagers I interviewed waited with great anticipation for the weekly call from their father or mother, and other schoolchildren discussed how they saved their lunch money just to call their parents and hear their voices. Both parents and children agreed that increased and regular communication facilitated more involvement in children's academic and emotional growth. One 14-year-old schoolgirl talked about how she called her mother for encouragement before her examinations. Another 16-year-old schoolgirl noted that, while she lived close to her father and he was financially supportive, she felt more comfortable asking her mother who lived in the Cayman Islands for money "for me to buy my sanitary napkins, body products, girls stuff." Young men noted the importance of their (physically) absentee fathers and the importance of their fathers as potential role models as they develop a sense of masculinity and make their transition from childhood to manhood (Chevannes 2001). A 17-year-old young man named Jesse in rural Jamaica found the phone to be critical in negotiating his estranged relationship with his father. Prior to owning his own mobile phone, he could not easily contact his father without asking to use someone's landline or asking for money for a phone card to initiate the call. When he asked for the money, his mother's family (who he lived with) would always ask why he needed the phone call or who he was calling and his father claimed he was reluctant to call the family home number because he had not maintained good relationships with Jesse's mother and family. With the cell phone, conversations could occur for the first time away from the watchful eye of his mother's family and the relationship had developed to such an extent that his father was helping him to apply for a place at a community college in Miami and was prepared to finance his studies.

Adult children living in the U.K. and U.S. also noted that they were keen to purchase mobile phones for their parents in order to keep abreast of their health and wellbeing; those whose parents lived more remotely noted that this was often the first time they could conceive of regular contact (Wilding 2006). In many instances this meant that family members contributed direct financial or medical help. When the Burke family's mother, "Mama", experienced a stroke, her children caring for her in Jamaica were financially strapped and could not pool enough money to pay for her medication. However, they did manage to gather enough money to purchase a J$240 phone card to call their brother living in the States to tell him about their mother's condition. Within a day, their brother sent the critical US$50 (J$3000) they needed for her medication. Indeed, some of the most dramatic implications of the mobile for transnational communication revolve around the fact that mobile phones have enabled emigrant parents and children with relatives in Jamaica to participate in the day-to-day affairs of their family members (Olwig 1999; Soto 1987).

MOTHERS AND GRANDMOTHERS IN THE TRANSNATIONAL FAMILY

The extensive literature on gender and family in the Caribbean often stresses the important role of mothers in Caribbean society: from Edith Clarke's famous book *My Mother Who Fathered Me* (1999), to Raymond T. Smith's discussion of the matrifocal family (1995), to more recent work on popular dancehall culture that emphasizes the often romantic depictions of mothers (Cooper 2004). Mothers and grandmothers play the most central roles in the family and household unit. Indeed, mothers and grandmothers have always played a central role in childcare, often facilitating their children's ability to take advantage of educational and occupational opportunities on a temporary or permanent basis, reinforcing the key role of mothers and grandmothers in the family. For those migrants who return, Plaza (2000) suggests that this central female figure of the household is also prevalent among Caribbean migrant communities and notes the emergence of "transnational grannies" who travel between the U.S., Canada, the U.K. and the Caribbean to visit, while also bearing food, gifts and other household items associated with Jamaican culture. However, while travelling is an ideal, for many returned migrants this ideal remains difficult to realize. For example, during a visit to Jamaica in 2007, Mrs. D, one of the participants in my previous study of returned residents, received some very distressing news. Her son and his wife had taken her teenaged grandson for a doctor's visit for testing. A few days later they learned that he had acquired a serious form of cancer and that her grandson would spend the holidays undergoing chemotherapy. As a nurse and a grandmother separated from her children and grandchildren living in London, her first instinct was to begin booking a flight to London. She had, after all, looked after some of her grandchildren while their parents were working and, at the very least, she could help out by cooking or driving her grandson to and from the hospital. Despite her desire, there were other matters to consider. Her husband had a medical condition and could not take the long, nine-hour flight to London from Jamaica and flying to London without him would mean that she would be leaving one worry behind for another. In the end, she stayed in Jamaica, relying on her eldest daughter in London to take on the caregiver role and to keep her abreast of the latest news about her grandson and her son's emotional state. This meant calls needed to be made at least a few times per week and very quickly she began weighing her options as to the most reliable and cost-effective phone plan—a J$1,000 pre-paid phone card for one thousand international minutes—which she began religiously purchasing and using at the beginning of each month. She reserved her landline for the receipt of local and international calls and often returned calls on her mobile. She now talks to her grandson directly when he is home and can report to others his progress based on his tone of voice and ability to carry on a conversation. The ability to hear sounds, background noises and changes in tone also gave Mrs. D an ambient sense of presence, despite the distance.

For many returned migrant grandmothers, migration or return to Jamaica to retire incites feelings of loss and ambivalence (Bauer and Thompson 2006; Horst, forthcoming; Olwig 2006). On the one hand, migration to the U.S., United Kingdom and elsewhere created financial independence and security for women, which provided them with the ability to fiscally supplement the historical importance of mothers and mothering in the Caribbean family. Many returnee women in my study of return migrants noted that the return to Jamaica, often prompted by their husbands (Chamberlain 1997; Horst 2008), was harder than they imagined, with a number of returnee women feeling they had abandoned their children and abdicated their role as the mother of the family by returning. This was compounded by the fact that a return to the Caribbean often meant the return to more traditional Caribbean gender roles, even for men who took on larger roles in the domestic setting as cooks and caretakers for the family while living in England. Many returnee mothers' and grandmothers' proficiency in navigating the price structure of mobile phone plans and cards in the name of transnational communication, care and (grand)mothering became a way to counteract the distance and ambivalence felt about their return and role in the family.

GENDER, POWER AND TRANSNATIONAL RELATIONSHIPS

Mobile phones have made it possible for children, caretakers and parents to gain access to regular communication with their children, grandchildren and parents. As discussed in the previous section, many Jamaican returnee's struggle with the loss of sense of self as a mother and grandmother once they have returned, particularly given their return to middle class norms surrounding gender and domestic work after a period of more equitable division of labor in England.[2] Whereas the mobile phone mitigates, to some extent, the distance of communication, social and structural distance associated with living abroad, acquiring a visa or green card and access to other resources remains another matter altogether. Such gendered geographies of power are particularly marked for women involved in relationships with Jamaican or foreign men living off the island. Lisa, a 32-year-old living in rural Jamaica, maintained a relationship with her overseas boyfriend through intermittent phone calls as well as through letters and small packages. One day she showed me a box of gifts that included a small photo album with copies of pictures she had developed for Robert, her Jamaican-born boyfriend living in Germany, as well as digital photos on her phone they took during Robert's last visit. A considerable number of Lisa's photographs were of the house and life of her boyfriend in Germany and the life they imagined together. In the album, Lisa had inserted pictures of Robert as a younger man visiting Jamaica, Robert's children, what would be Lisa's car, as well as pictures of the kitchen, bedrooms and other rooms of the house they planned to share. Lisa admitted that the mobile phone played an important role in developing her relationship with Robert because they talked and sent text messages several times a week. Robert and Lisa even discussed marriage. In fact, everything seemed to be going well until a friend of Lisa became jealous and started to spread rumors about Lisa, which made their way back to Robert in Germany. Robert started to call Lisa several times each day, harassing her about her whereabouts, asking about background noises, what she was doing and how she was raising her son. Shortly thereafter, he stopped sending money for her upkeep and phoned to say that her application to join him in Germany was denied.Surveillance, often related to accusations of infidelity, also represented an important and often unanticipated side effect of increased communication. By being able to call at any hour of the day to harass about whereabouts, the noises in the background and the minutiae of everyday life, the mobile phone began to feel like a form of surveillance rather than enhanced communication. While the mobile phone facilitated an increased ability to communicate and extend one's networks outside Jamaica, it ultimately did not dramatically change the gendered power dynamics that underpinned their relationship. When Lisa's boyfriend Robert decided that he did not like the way she responded to him on the phone or did not like hearing a male voice in the background, he had the power to withdraw his financial and emotional support.

Lisa was clearly disappointed with how things turned out after she invested so much time and energy in her relationship, but Lisa was not naïve about the prospects of the relationship, her tenuous access to mobility and the economic asymmetries between them. She had not, for example, quite ended her relationship with her Jamaican boyfriend in Jamaica while she was involved with Robert. She also kept her eye on opportunities for modeling which she viewed as a potential career path. In essence, Lisa utilized her mobile phone in her efforts to cultivate her identity and aspirations for mobility in the context of uneven access to economic and social resources that reflect a broader power (and gender) dynamic reflected in transnational migration and mobility (Batson-Savage 2007). One of the key differences between Lisa's relationship with Robert and Winston and Donna's relationship is that Donna possesses the economic advantage, mobility and status as a citizen of the United States and, importantly, Winston stays in Jamaica by choice, a factor that removes an element of the asymmetry between them.

The gendered geographies of power also influenced the ways in which young men in Jamaica viewed mobile phones and the ability to maintain transnational relationships with family members and others living outside of Jamaica. "Indian", a 20-year-old man who sold hard candy and nuts on the roadside in rural Jamaica, viewed the mobile phone as a way to maintain a connection with his "links" (connections) outside of Jamaica. Stressing the difficulties of life as a young man in Jamaica where unemployment hovers around 13%, Indian was particularly perceptive of inequity between the opportunities available in Jamaica and abroad, and expressed how he felt it was almost impossible to "move forward in life" in Jamaica supporting a girlfriend and young child without such connections. Moreover, and like others, he believed that Jamaicans living "in foreign" possessed an obligation to take advantage of their opportunities to support their kin who have not been so fortunate. His primary role was, in effect, to communicate "the situation out here".

Although he admitted to calling abroad on a regular basis, Indian's relatives tended to send money to his grandmother who redistributed it within the family as she saw fit. Not only did family members abroad see her as the center of the family, they also believed that she was no longer able to make money herself; young people like Indian could always, from their perspective, "find a work". This ultimately meant that Indian received relatively little in the way of direct contributions. Over time Indian began to resent his lowly position within the extended family as well as the attitude that, as a young man, he must be able to find work; women and the older generation, he felt, were more likely to garner the sympathies for their situation. In response, Indian decided to more actively cultivate relationships with his male "links" from the community, high school and extended family now living in the U.S. About every two weeks, Indian called his relatives living in Brooklyn and New Jersey, links that might result in US$20 or US$30, a sizeable portion of his monthly income. While these amounts are small compared with what is sent to his grandmother, they are sent directly to him from his brother and other cousins who are more sympathetic to his struggle. In exchange for "a smalls" (change or a small amount of money), Indian leverages his ability to keep the links of communication open and keep them up-to-date on the local happenings and music scene. When his relatives return to Jamaica to visit at Christmas or other holidays, Indian locates a local mobile phone or SIM card for them to use, arranges for a car or taxi driver, locates mangoes, ackee, coconuts and other Jamaican delicacies or takes them to dancehall sessions, shows, bars and other places where they can experience being home in Jamaica. Most of these activities will still be funded by the visiting relations and friends given Indian's tenuous economic situation, but this process of keeping people abroad connected to authentic Jamaican culture while they are home—tasting ackee, jerk pork, smoking ganja, drinking overproof rum and so on—facilitates the experience of coming home and makes those visiting from abroad return home with style and status, "like a big man". This, in turn, provides Indian with his own sense of status and opportunity, transforming Indian into a "big man" during these visits. The mobile phone remains one of the key ways to keep his links, or connections. In his case, this helps him to subvert the gendered and generational hierarchy associated with the remittance economy in many Jamaican households.

CONCLUSION

Mobile phones, video cassettes, e-mail and social network sites have dramatically altered the frequency and meaning of communication for transnational families, creating spaces through which family members can connect and continue to build a sense of family (Goldring 1998; Smith and Guarnizo 1998; Mahler 2001; Richman 2002; Uy-Tiocco 2007). Yet new communication technologies such as mobile phones are more than just the "means" of communication. As Vertovec (2004, 15) has argued for, phone cards, mobile phones and other mundane communication technologies "are increasingly used transnationally to link migrants and homelands in ways that are deeply meaningful to people on both ends of the line." For many transnational families, the availability and ownership of mobile phones has helped to collapse, at least temporarily, the distance between Jamaicans at home and abroad due to their ability to create a sense of involvement in each other's everyday lives. It has also enabled Jamaicans at "home" to communicate their care and concern for their friends and family "in foreign".

While acknowledging that the mobile phone is only one object in an ever-evolving palette of digital media and communication technologies, I focus upon the mobile phone in order to illustrate the particular impact of one communication technology in shaping how Jamaicans express and communicate across transnational space. The mobile phone became central to making specific requests to include special items in the "barrels", such as shoes and clothing, appliances, soap and detergents, as well as basic foodstuffs such as rice and cooking oil. It was easier to be involved in the day-to-day events of someone's life abroad and maintain romantic connections. Grandmothers integrate the cell phone into the care work they continue to do across national borders and young Jamaican men may make a brief "link-up" that, in essence, cultivates a male kinship and friendship network across national borders. These activities provide many Jamaicans with the opportunity to become active agents within, and possibly transform, transnational social spaces in a fashion that they were not afforded when the power to initiate communication depended on the inclinations and benevolence of others. Yet they also reveal the gendered geographies of power in transnational spaces wherein women in Jamaica continue to leverage their sexuality and role as mothers or caregivers and young men leverage their access to networks, goods and Jamaican culture as they look for opportunities to "link up" and stay connected, key practices to realize their aspiration to move forward in life.

ACKNOWLEDGEMENTS

An earlier version of this article was published in *Global Networks: A Journal of Transnational Affairs*:

Horst, Heather A. 2006. The Blessings and Burdens of Communication: The Cell Phone in Jamaican Transnational Social Fields. In *Global Networks: A Journal of Transnational Affairs 6(2):* 143–159.

I am grateful to Global Networks for the opportunity to publish portions of the article.

NOTES

1. These services have increased significantly in the past five years. It is now possible to add credit through "credit you-credit me" services and for Jamaicans living abroad to add value to another Jamaican's phone (or their own phone) through a web browser.
2. This is not to suggest that in England all men assumed roles as caretakers of the household and children, but most returnees I interviewed noted that the conditions of work and the necessity of women working as nurses at night meant that men had to assume more domestic responsibility. These responsibilities diminished as children grew and, unless there was a task a husband particularly enjoyed, often disappeared upon their return to Jamaica.

REFERENCES

Basch, Linda, Nina Glick Schiller, and Cristina Szanton-Blanc. 1994. *Nations unbound: Transnational projects, postcolonial predicaments and deterritorialized nation-states.* Amsterdam: Overseas Publishers Association.

Batson-Savage, Tanya. 2007. "Hol' awn mek a answer mi cellular": Sex, sexuality and the cellular phone in urban Jamaica. *Continuum* 21(2):39–251.

Bauer, Elaine, and Paul Thompson. 2006. *Jamaican hands across the Atlantic.* Kingston: Ian Randle.

Boyett, Inger, and Graeme Currie. 2004. Middle managers moulding international strategy: An Irish start-up in Jamaican telecoms. *Long Range Planning* 37:51–66.

Brown, Dennis A. 1997. Workforce losses and return migration to the Caribbean: A case study of Jamaican nurses. In *Caribbean circuits,* ed. Patricia R. Pessar, 197–223. New York: Center for Migration Studies.

Chamberlain, Mary. 1997. 1997. *Narratives of Exile and Return,* First edition. Oxford: Macmillan/ New York: St. Martin's Press.

Chevannes, Barry. 2001. *Learning to be a man: Culture, socialization and gender identity in five Caribbean communities.* Mona: University of the West Indies Press.

Clarke, Edith. 1999. *My Mother Who Fathered Me: A Study of the Families in Three Selected Communities of Jamaica.* Kingston: University of the West Indies Press.

Cooper, Carolyn. 2004. *Sound clash: Jamaican dancehall culture at large.* Oxford: Palgrave Macmillan.

Foner, Nancy. 1983. *Jamaican migrants: A comparative analysis of the New York and London experience.* New York: Center for Latin American Studies.

———. 2005. *In a new land: A comparative view of immigration.* New York: New York University Press.

Fortunati, Leopoldina, and Jane Vincent. 2009. Introduction. In *Electronic emotion: The mediation of emotion via information and communication technologies*, ed. Jane Vincent and Leopoldina Fortunati, 1–31. Oxford: Peter Lang.

Goldring, Luin. 1998. The power of status in transnational social fields. In *Transnationalism from below*, ed. Michael P. Smith and Luis E. Guarnizo, 165–195. New Brunswick, NJ: Transaction Publishers.

Goulbourne, Harry, and Mary Chamberlain, eds. 2001. *Caribbean families in Britain and the transatlantic world.* London and Oxford: Macmillan.

Hochschild, Arlie. 2000. Global care chains and emotional surplus value. In *On the edge: Living with global capitalism*, ed. William Hutton and Anthony Giddens, 130–146. London: Jonathan Cape.

Horst, Heather A. Forthcoming. Reclaiming place: The architecture of home, family and migration. In *Anthropologica*.

———. 2009. Families. In *Hanging out, messing around, and geeking out: Kids living and learning with new media*, ed. Mizuko Ito et al., 149–194. Cambridge, MA: MIT Press.

Horst, Heather A., and Daniel Miller. 2005. From kinship to link-up: The cell phone and social networking in Jamaica. *Current Anthropology* 46:755–778.

———. 2006. *The cell phone: An anthropology of communication.* Oxford: Berg.

———. 2008. Landscaping Englishness: Respectability and Returnees in Mandeville, Jamaica. *Caribbean Review of Gender Studies* 2(2), 2008.

Ito, Mizuko, Daisuke Okabe, and Misa Matsuda. 2005. *Personal, portable, pedestria: Mobile phones in Japanese life.* Cambridge, MA: MIT Press.

Levitt, Peggy, and Nina Glick Schiller. 2003. Transnational perspectives on migration: Conceptualizing simultaneity. Working Paper 3–09J, Princeton University Center for Migration and Development. http://www.peggylevitt.org/pdfs/cncptualzng_smltaneity.pdf.

Ling, Richard, and Brigitte Yttri. 2002. Hyper-coordination via mobile phones in Norway. In *Perpetual contact: Mobile communication, private talk, public performance*, ed. James Katz and Mark Aakhus, 170–192. Cambridge: Cambridge University Press.

Mahler, Sarah J. 1998. Theoretical and empirical contributions toward a research agenda for transnationalism. In *Transnationalism from below*, ed. Michael P. Smith and Luis E. Guarnizo, 64–102. New Brunswick, NJ: Transaction Publishers.

———. 2001. Transnational relationships: The struggle to communicate across borders. *Identities* 7:583–619.

Mahler, Sarah J., and Patricia Pessar. 2001. Gendered geographies of power: Analyzing gender across transnational spaces. *Identities* 7:441–459.

Miller, Daniel. 2009. *Anthropology and individuals: A material culture approach.* Oxford and New York: Berg.

Olwig, Karen F. 1999. Narratives of the children left behind: Home and identity in globalised Caribbean families. *Journal of Ethnic and Migration Studies* 25:267–287.

———. 2006. *Caribbean journeys: An ethnography of migration and home in three family networks.* Durham, NC: Duke University Press.

Parreñas, Rhacel S. 2005. *Children of global migration: Transnational families and gendered woes*. Stanford, CA: Stanford University Press.

Pessar, Patricia, and Sarah J. Mahler. 2003. Transnational migration: Bringing gender. *International Migration Review* 37(3):812–846. Published by the Center for Migration.

Plaza, Dwaine. 2000. Transnational grannies: The changing family responsibilities of elderly African Caribbean-born women resident in Britain. *Social Indicators Research* 51(1):75–105.

Richman, Karen. 2002. Miami money and the home gal. *Anthropology and Humanism* 27(2):119–132.

Smith, Michael P., and Luis E. Guarnizo, eds. 1998. *Transnationalism from below*. New Brunswick, NJ: Transaction Publishers.

Smith, Raymond T. 1995. *The Matrifocal Family: Power, Pluralism and Politics*. London: Routledge.

Soto, Isa M. 1987. West Indian child fostering: Its role in migrant exchanges. In *Caribbean life in New York City*, ed. Constance R. Sutton and Elsa M. Chaney, 121–137. New York: Center for Migration Studies.

Thomas-Hope, Elizabeth. 1988. Caribbean skilled international migration and the transnational household. *Geoforum* 19:423–432.

———. 1992. *Explanation in Caribbean migration*. London, Macmillan.

Thompson, Paul, and Elaine Bauer. 2000. Jamaican transnational families: Points of pain and resources of resilience. *Wadabagei: A Journal of the Caribbean and Its Diaspora* 3(2):1–36.

Turkle, Sherry. 1984. *The second Self: Computers and the human spirit*. New York: Simon and Schuster.

Uy-Tioco, Cecilia. 2007. Overseas Filipino workers and text messaging: Reinventing transnational mothering. *Continuum* 21(2):253–265.

Vertovec, Steven. 2004. Cheap calls: The social glue of migrant transnationalism. *Global Networks* 4:219–224.

Vickerman, Milton. 1999. *Crosscurrents: West Indian immigrants and race*. Oxford: Oxford University Press.

Wajcman, Judy. 1991. *Feminism confronts technology*. University Park, PA: The Pennsylvania State University Press.

———. 2004. *Technofeminism*. Cambridge: Polity Press.

Wilding, Raelene.2006. "Virtual" intimacies? Families communicating across transnational contexts. *Global Networks: A Journal of Transnational Affairs* 6(2):125–142.

5 Mobiles, Men and Migration
Mobile Communication and Everyday Multiculturalism in Australia

Clifton Evers and Gerard Goggin

INTRODUCTION

There is a wide recognition of the importance of media—especially new information and communications technologies—to the processes of migration, resettlement and cultural negotiation. Existing research suggests that, especially for new migrants, mobile communication plays an important role in how they negotiate new social dynamics. However, we are still at early stages of understanding the deeper cultural and embodied dimensions of this process.

In this paper we take a broadly cultural approach to the topic of gender, migration and mobiles. By "cultural" we mean this work draws upon cultural theory and cultural and media studies (Goggin 2008a, 2008b; Hjorth 2009). We are especially interested in the way that mobile media—text, pictures, videos, music and affordance of mobile phones—harness the "cultural, class, ethnic, gender, and sexuality discourses [people] already perform in face-to-face interactions" (Prøitz 2005). What we wish to explore is the intersection between masculinity, migration and mobiles, especially to give attention to the role played by affects and embodiment. Our approach draws on fieldwork conducted in suburbs in inner and outer Sydney, with both new and more established migrants. The chapter aims to contribute to the emerging literature on migration, diaspora and technology by offering new insights into the mobile and how cultural diversity plays out—but also how masculinity is negotiated within this milieu.

MIGRATION AND MULTICULTURALISM IN AUSTRALIA

Migration has been a decisive factor in the constitution of contemporary Australian society, especially since World War II (Jupp 2007). Flows of migrants from Asia-Pacific countries in the 1970s through to the 1990s have been especially important in cementing Australia's place in that region, rather than through the colonial period (19th century) and post-Federation period (first half of the 20th century) imagining itself as part of England

and the British dominion. Since the late 1990s, migration to Australia has steadily continued, despite the controversial crackdown on refugees, especially those arriving by boat (Marr and Wilkinson 2003). Extensive international migration feeds into Sydney's growth, and there are groups with high humanitarian visa distribution including people from Congo, Liberia, Cambodia, Laos, Burundi, Nigeria, Sudan, Iran, Bosnia, Macedonia, Serbia, East Timor, Iraq, Afghanistan, as well as Assyrians, Kurds and Tamils. In the 2006 census 37% of the total Sydney population was overseas-born. The area in Sydney's west has the highest proportion of people born overseas (Australian Bureau of Statistics 2008).

We both come to this topic of men's migrations and the use they make of and meanings they attach to mobiles through ongoing fieldwork in the area. Goggin is undertaking a collaborative national study on youth and mobile media (described in Crawford and Goggin, forthcoming), one of the sites of which is in the inner-west of Sydney, home to a mix of established Anglo, Greek, Portuguese and Italian communities, arrivals in the 1970s and 1980s from Vietnam and China particularly, then-recent arrivals from African countries and elsewhere. Evers has worked for some time on numerous projects with young refugee men, particularly from Africa. We will draw on observations and data from both these projects, but will especially focus on the fieldwork Evers has undertaken with young male soccer players in outer-west Sydney, and part of a refugee resettlement assistance sport program called "Football United".

STUDYING AFFECT AND MIGRANT MEN

The Football United program assists recently arrived humanitarian refugee young people and their families during their transition into Australian society. A shared love of football (soccer) is used to present opportunities for belonging, mentoring, community integration and sharing cultural capital and resources.

Many of the young people and families that take part in Football United have fled from civil conflicts and have experienced the trauma of loss, separation, as well as family and community breakdown (Thorell 2007). As well as the multiple issues of adjusting to a new country, "Refugee young people must also negotiate family, peer, individual and community expectations within the context of adolescence" (Olliff 2008, 53). Many of the young men who participate in Football United are from various ethnic and regional affiliations from Sudan: Dinka, Nuba, Nuer and Achole. Twenty-eight percent of Sudanese-born migrants live in Sydney (Australian Bureau of Statistics 2009). Many of the young women from the same communities experience cultural barriers to participation in sport, and favor forming friendships with the women in the program when they do have the opportunity to play football. However, in this paper it is the young migrant men and the construction of masculinity through mobiles that is the focus—for a number of reasons.

We are certainly mindful of the importance of relationships among men and women, and also the inextricability and co-definition of masculinities and femininities when it comes to ICTs and migration. However, the ethnography research drawn upon here centered on participation-observation ("hanging out") with young men. More importantly, the opportunity afforded was to cast light upon the neglected area of masculinities and mobile phones where specific conditions of possibility play out. Research on mobiles has included some significant work on gender, exploring its construction through technology use, discourses and social relations in which both women and men are differentially or commonly involved. There are also a number of studies that take femininity as their focus. However, there has been little work to date that explicitly investigates and theorizes masculinity and mobiles—let alone in the processes of migration and diaspora.

These young men who are the focus of this study negotiate social expectations of being an adolescent, doing gender and resettling through their mobile phone use and concomitant representations. Our fieldwork acknowledges that experiences of gender and mobile phones are contingent events. Let us be clear: we are not looking for a pattern to masculine interaction with mobile phones during migration but rather we seek to map the particular experiences of masculinity and mobile phones during migration for these young Sudanese men. This means that our approach involves participating in "patterns of movement while conducting ethnographic research" (Sheller and Urry 2006, 217; cf. Höflich and Hartmann 2006). This mobile fieldwork involves mapping phone use and practices as we "hang out" with the young men. As Awad Ibrahim (2008) stresses in his study of migrant African youth in Canada, hanging out accommodates how young people's "identity is multiple and performed in multiple ways and sites" (244), as well as providing the opportunity to note verbal and non-verbal performance.

Such an approach is important because it taps into how we understand that to *do* masculinity is to negotiate complex and contingent experiences where space, psychology, biology and the social intermingle and shift. We accommodate Judith Butler's (1990) argument that all gender is performative, "the repeated stylisation of the body" (33, 136). Repetition and recitation establish some dispositions as "masculine" and others as "feminine". Further, the way young men make sense of and do masculinity is as an expression of historical, social, biological and cultural exigencies, involving what Pierre Bourdieu called "cultural capital" (1984).

Bodies soak this cultural capital into their skin and a gendered "habitus" develops with which they exercise their cultural capital through the way they conduct themselves and come to see the world—their disposition. A "masculine" disposition gets deep into the bodies of the young men and, along with other discourses on ethnicity, migration, class and the like, influence what the young men do: the way they walk, how they speak, the way they gesture, their taste for certain foods and fashion and so on. While the young men's habitus can be quite resistant to change, because bodies find it hard to forget deeply embodied discourses, they can and do negotiate new cultural capital and tastes in regards to doing masculinity in Australia. Mobile phones play a significant role in this negotiation. For example, as we shall show, through mobile phones the young men can display their taste and test what might be appropriate or not in Australia. They build up networks that reaffirm their own tastes borne of migration and their homeland, as well as provide a resource for accessing new cultural capital during resettlement. Young men's emotional and affective lives underpin this process and influence what they do and how they do it. There is a sensual life to masculinity (Seidler 2007; Lilleaas 2007; Evers 2004, 2006; Connell 2002, 2005) that motivates learning and adaptations during resettlement. Contrary to popular assumption young men use feelings to navigate and make sense of the world around them, which involve bodies affecting others and being affected (Tomkins 1962–1992).

While affect theory is well-established in cultural theory, as yet we have few studies of affects and mobiles. There has been some pioneering work on emotion by Jane Vincent (2005) and Leopoldina Fortunati (2005) and not least in their 2009 collection (Vincent and Fortunati 2009), and there are important contributions from other scholars, notably Raul Pertierra in his work on intimacy (2005). While broadly related, there are important differences in the concept of affect, there is little work on this with regard to mobiles (exceptions include Lasen's early 2005 paper and Crawford 2009). We are interested in how cultural studies work on affect, deriving from Silvan Tomkins (1962–1992), developed by Eve Kosofsky Sedgwick (1995), Elspeth Probyn (2005) and others. For our purposes here, we draw on Anna Gibbs's (2001) discussion of the relationship between media and affects, where she sees media as an amplifier, heightener and intensifier of affects and emotions. For example, affect can exist within the text itself, and arise from a page or screen as it is read, and is capable of affecting us in ways that stimulate, arouse, exhaust, inspire, disgust, humiliate, distress, startle, stress, anger, thrill and the like. Further, media cannot be understood independently of its material form and the affective relationships this builds. We could regard different models and features of phones and how they variously enhance, highlight or diffuse different affective mediation, such as through video and camera features, the layout of the buttons, screen operation and the material form of the phone. Mobile phones are not separate to bodies but part of them as affects leap from body to body through phone mediation. The importance of reworking mobile phones in terms of affectivity lies in the way it draws attention to the richness of bodies feeling, learning and doing during experiences involving mobile phones.

NEW ARRIVALS AND MEN'S CULTURES

The two leading studies of mobile and migrants in Australia have focused on the controversial topic of refugees. Linda Leung explores how the mobile phone is "appropriated by refugees in immigration detention" (2007, 1). Crucially, she finds, the mobile phone helps refugees sustain "connection with their imagined communities" (Leung 2007, 3) while behind barbed wire and at the desert-based detention centers. Diana Glazebrook (2004) has studied Hazara refugees with temporary protection visas (so-called "TPVs") upon release from immigration detention centers (2004, 52). Use of mobile phones is seen as a priority for Hazara refugees because it helps them to be both resourceful and tactful in managing their resettlement.

The practical, functional qualities of mobile phones for negotiating bureaucratic resettlement discerned in these previous studies of Leung (2007) and Glazebrook (2004) are present in the young men participating in the Football United program. A common post-arrival barrier for the young men of Football United is employment: unfamiliar work cultures, lack of fluency in the language of the resettlement country, qualifications not recognized and lack of relevant networks. As such the young men tend to gain employment in low-status and low-paying positions, such as factory-based shift work. The young men move between work, English classes, football, friend's houses, migrant resource centers and the like. A mobile phone becomes a necessity to stay in touch with them as they are always on the move. The importance of being a male in the young men's cultures is what gives them this independence to come and go as they see fit, and so influences their mobile phone use and ownership.

Because the young men are believed to be the "future" of the family, they are the first to really begin interacting and making their way through the broader community. They tend to be more adaptable to the new culture as their cultural capital and habitus is less sedentary and settled, while the parents' habitus is far more ingrained so they feel "lost", intimidated or are more cautious as they have experienced so much hardship. The young men also are familiar with that cross-cultural masculine expectation of needing to "conquer challenges" and "be tough". At the onset of puberty many of the young men are considered "adults", the concepts of "youth" or "adolescence" are foreign to the families and cultures. Many of the young men involved in Football United must combine participation with employment, schooling, domestic family responsibilities and an expectation that they help extended family members remaining in Africa (Gow 2002; Kirk and Cassity 2007).

However, the culture of Australian men in sporting environments— whom the refugee young men must negotiate with because they dominate the resources of public space, resources and facilities—can be confronting, often characterized by a "culture of drinking alcohol after/during games, swearing, sledging or aggressive competition" (Olliff 2008, 58). When the families get to Australia the first thing they spend the government money on tends to be mobile phones; these are distributed to the young men first and foremost. "M" spoke to Evers about how he could afford to buy a phone when he arrived in Australia:

> In my country we don't have the money to get it [phone], like it is too expensive and rules in the country there are rations . . . I saw a soldier holding his gun in his hands and walking and he take a phone off my friend. (Interview, March 2009)

A recurrent difficulty for these young men is being able to afford to use their mobiles. Some of the young men have landlines, but the lines are cut off intermittently as their families struggle to pay the bills. Internet connection and the concomitant e-mail contact are also sporadic for the same reason. As such, for the young men, their mobile phones are their communication technology of choice because they can control payment of the bills.

The young men all talk about how the mobile phone was given to them despite costs involved because they go out into the new culture and spaces and have to report back regularly. There are heightened concerns for their safety in the Australian culture, giving their important role in trail-blazing for their family. This "safety" discourse attached to the mobile phone means the young men use their phones to reassure their parents, map where they have been and find support when they need it. This is in marked difference to Australian young men who do not bring up such "safety" reasons for mobile phone ownership, justification and use—whereas young Australian women do.

While mobile phones can contribute to resourcefulness and safety—read as independence—while resettling it can also cause problems. "N" told Evers that for his community, "independence" can appear more like "rebellion" and "cultural loss" to the older generation who want to hang on to community solidarity and need their young men to be "team players". N explained that while some of his friends use their phones to "get away" from their families and parent culture, others embrace their "roots" and "responsibilities", while most are somewhere in-between. The mobile phone presents a collection of opportunities but also a collection of responsibilities that require nuanced and ongoing negotiation.

MORAL PANICS

A lot of the phone credit the young men of Football United get for football coordination and for "official" settlement requirements actually goes to downloading music. Indeed the young men prefer phones that have the capability to play music. Phones without cameras, music capabilities and Bluetooth are sold or traded away in what has become a black market of mobile phones.

˙   At the football field the boys would sit around in a group and Bluetooth each other music tracks. Bluetooth is used because it is free to transfer files, building a music library. The phones always have music blaring out of them. The boys cannot afford MP3 players and the phones serve this purpose. The music is generally played out loud rather than through head-phones. Of particular note are the boys who bring back phones from African countries after they return from a visit with families and friends. The phones in Sudan have much louder volume capabilities. M told Evers, "They [Sudan] have mad phones, they are good phones. They have a louder speaker . . . In Australia all the phones they have small speaker but in my country, the phones with the speaker on them are louder" (Interview, March 2009). While the music enables the young men to bond with others over taste, it also marks out an aural territory, especially in public spaces (Bull 2008; Evers 2008)—something valuable because at home privacy is hard to come by due to large families and small residences. This marking out of an aural territory through their mobile phones' music capabilities puts the young men at odds with other people on public transport or in public areas who view such an incursion into "their space" as "disrespectful", "impolite" and "anti-social".

This discourse of the young men being "anti-social" perpetuates a stereotype of them as "threatening, at risk, vulnerable and in need of control, curbing, fixing, developing or cultivating" (Westoby and Ingamells 2007, 54). This is particularly the case in regards to young refugee men, who have to deal with a public discourse that constructs young people as a disengaged group and "at risk" of "becoming problems" (even "dangerous") (Couch 2007, 40). In Australia young African men have been framed as gang members and sexual predators, and have been the subject of moral panics (Cunneen 2007; Goggin 2010; Kerbaj 2006; Leach and Mansouri 2003). More often than not the groups of men referred to as "gangs" are simple friendship groups. Many of the young men don't even have their family here and have moved with their mates through refugee camps since they were born. Relatives have been killed in the wars in Africa or are stuck in over-crowded refugee camps. Not surprisingly, the boys hang out together in tight-knit groups and back each other up if they feel threatened. Ethnic profiling and criminalizing of non-white groups of men creates an "us" and "them" mentality that reaffirms Anglo privilege in Australia (Collins et al. 2000; Poynting and Morgan 2007).

The effects of the negative discourse on young men from Africa are very real. Out front of shopping malls security guards send the boys away for indiscretions like playing their music too loud on their mobile phones. Often security guards target the wrong boy and attribute "offences" to him because all they see is "black". Other members of the public are hostile to the young men and complain about them hanging out, talking loudly on their phones and their music. Confident body posture and belief in their own cultural ways of doing things transgress the "normal" protocols around mobile phone and personal space established by locals. Ironically, the mobile phone for these young men is not about intimidation or rebellion but about marking out some public space where they can feel at home and safe in a strange place. It could also be easily seen as everyday innovation and cultural creativity of these emergent and still little-understood youth cultures (Butcher and Thomas 2003; Goggin 2010).Feeling Mobile Phones For the young men mobile phones are part of their sensual life. With a mobile phone there is the possibility of contacting family and friends in their homeland any time. However, sometimes this contact comes at an emotional cost. As Glazebrook (2004, 50) explains in regards to newly released Hazara refugees, "Sadali avoids phoning his family because it takes him a week after phoning them to recover from his sense of helplessness and loss in their absence."

While there is such emotional and affective contact with the family there is also emotional and affective exchange with other young men here in Australia. Mobile phones are part of a masculine bonding that is crucial to coping with resettlement. The group of young men involved with Football United have a passion for their mobile phones that goes beyond simply need. The young men are intimately connected to their phones. They compare their phones, play with them, swap them, compare and share what's on them and always guard them with vigor. "S" told Evers, "If you have a good phone everyone will like to play with it, everyone will like to touch it, but if you have a crappy phone, no one likes it and no one cares about it . . . it gets you respect" (Interview, March 2009). Some boys will take pictures of girls and pass them around, but they also take photos of their own bodies to share; as "B" explains, 'They take pictures like taking their clothes off, showing the abs and all that" (Interview, March 2009). When playing football some of the young men will not put their phones down and will carry the phone in their hand, music on. The affectivity phones enable also helps fuel affective bonds between the young men.

The young men's bodies share affective and emotional memories of trauma, upheaval, as well as forced migration and resettlement. Their cultural and ethnic backgrounds also see them bond closely. The shared experience of a passion for football and the highs and lows of winning and losing adds to these emotional and affective and discursive bonds. Cultural capital is also refined and particular conversations, nicknames and handshakes emerge and are shared, and then mark out the group of young men as a friendship group. Support—financial, emotional, safety, cultural know-how—from others in the group is eventually forthcoming after a period of bonding and showing commitment to a particular disposition. This process of masculine bonding helps the young men make their way and cope during resettlement because it provides a way into the broader Australian community. To work their way into bonding with other men who hold power in the Australian community, the young men wear their love for, and skill at, football on their sleeve, the football cultural capital allowing them to mediate across ethnic and cultural divides. Some of the young men brand their phones with football team stickers and logos. This enables the first steps towards a sense of belonging in a culture that privileges sporting prowess links with masculinity.

Friendships are established that prove invaluable as resources during resettlement: the resources tend to build on the football cultural capital and slowly extend into other functional support for resettlement—employment contacts, directions, English instruction, transport and the like. However, through the phone the young men can not only *stay in touch* with the close-knit group of mates who are also refugees but also *get in touch* with powerful new mates in Australia, where men still receive significant "patriarchal dividends", if differentially distributed (Connell 2002, 142).

Through the phone the young men discuss with other young men not only football but sexual "conquests", women in general, football, cars, work, settlement issues, timetabling, families, politics, government, friendship and the like. They share pictures, movies, music and text messages. Text is the most popular medium, as "S" explained to Evers: "Mostly I use text because if I buy a $10 credit I get $70 text messages for one month and so I mostly use it for text" (Interview, March 2009). The young men use their phones to fuel and enable bonds with other men, and to facilitate the extension of their networks. Through the mobile phone the young men stimulate the senses, bodies and minds of other young men and so establish a connection. They can even "embody the text" through use of capital letters, emoticons and the lexicon of mobile messaging (Prøitz 2005). Sometimes the young men might circulate a picture of a girl they all think is "hot", a car they all agree they desire or footage of a football player's skills—which they then crowd around to watch while slapping hands, laughing or cringing while jostling to see.

The sharing of media via the mobile phone also provides an opportunity to explore taste. With the resettlement the young men must negotiate new masculine norms, tastes and dispositions to find which in this new place are "right" and "wrong". For example, some of the young men show pictures of their wives. The response from the dominant Anglo-Australian young men is shock. How can you have more than one wife? The affective reaction to the image lets the newcomers know that they have to be careful here, the codes and etiquette of relationships are very different in Australia from their own countries of origin. In contrast the Anglo-Australian men show some of the newcomers pornography on their phone to which some of the young Muslim African refugees reel back in shock.

This affective learning and sharing through the mobile phone is how cultural capital is negotiated, as can be shown through a particularly negative case. M saw footage of himself being beaten up during a football game. He was tired of fighting: in his home country he had been a child soldier. He promised to never fight again. Yet, the footage was circulated by mobile phones—and M was ridiculed by the local young men and his friends. His shoulders slumped and he kicked and stared at the ground. In a study of the role of shame Elspeth Probyn writes of how shame is used to set up how one is supposed to act (2005). There is shame experienced when people fail to perform because they are forced to confront their capabilities. In M's case, he fell into a few fights after this experience and was "red-carded" (that is, sent off the football field) numerous times as he tried to rebuild his "masculine" reputation and adopt the appropriate violent "masculine" disposition for fear of losing respect and being ridiculed again.

Our argument, then, is that there are significant interaffective experiences that contribute to how the young men learn to behave and do masculinity in Australia during resettlement. The registering of the media and mobile phones as an affective modality means that the young men learn through their bodies if something has changed, or needs to change. Their bodies shiver, heat up, cool down, agitate, cry, shake and the like when cultural capital and discourses deeply embedded in their bodies have been disturbed, or new conditions are unsuitable or challenging. When the young men first arrive in Australia, their proprioceptic, kinaesthetic and sensual awareness cannot settle even though there are all sorts of "support" in place. The young men slowly have to adapt to new movement, practices, social expectations, bodies and ecologies. The new conditions of possibilities force their way into the young men's habitus by way of sight, taste, smell, touch, hearing and balance. And one way this occurs is through their mobile phone use.

CONCLUSION

In some ways, our study of young African migrant men in Sydney, Australia confirms previous studies of the importance of mobiles in the lives of recent arrivals seeking to settle in a new society—while maintaining their links to and relationships with home cultures. What has become clear is that the mobile phone allows the young men to negotiate what are new ways of interacting and communicating and coordinating people, meeting, accessing resources and events. The mobile helps them configure their new mobility and opportunities for coming together across familial and social networks, and helps them organize flows of information, bodies and spaces. The mobile phones open up all sorts of sites, places, discourses and materialities. Some might see this as liberation, but it is not of choice (Ahmed 2004, 152).

Our study also suggests that mobiles are a strategically critical cultural technology for this group and, we suspect, for many other groups of migrants. Mobiles allow these young men to negotiate the conflicts, contradictions and possibilities of everyday life in contemporary Australia, while being intensely connected to familial and friendship structures and relationship, economic, personal and marriage projects in circuits of the

African diaspora. Mobiles go to the heart of this refashioning of culture in migration and globalization, represented here, for instance, by the importance of mobiles as an emblem (carrying them *while* playing soccer) or in the listening to music (what this represents for the individuals and their group, as well as how such performances are received, often negatively, by others in the tense public spheres of these cities).

What we also have been intrigued by in this study is the affective and bodily dimensions of this interplay between migrating men and their mobiles. How masculinity is performed and reconstructed necessitates intense affective investments, in which mobiles play a significant role in embedding these men in systems of gender, class, place and space. Much previous literature on mobiles and migration has focused on workers, and here we hope we have added a new emphasis—namely of the way popular cultural practices such as sport (often seen as simply "leisure" or "recreation") figure very prominently in the processes of migration, diaspora and how information and communications technologies are deeply implicated in these.

REFERENCES

Ahmed, Sara. 2004. *The cultural politics of emotion.* Edinburgh: Edinburgh University Press/Routledge.
Australian Bureau of Statistics (ABS). 2008. *Sydney: A social atlas.* Cat. no. 2030.1. Canberra: ABS. http://www.abs.gov.au/ausstats/abs@.nsf/mf/2030.1/.
———. 2009. *Perspectives on migrants.* Cat. no. 3416.0. Canberra: ABS. http://www.abs.gov.au/ausstats/abs@.nsf/mf/3416.0.
Bourdieu, Pierre. 1984. *Distinction: A social critique of the judgement of taste.* Trans. Richard Nice. London: Routledge and Kegan Paul.
Bull, Michael. 2008. *Sound moves: iPod culture and urban experience.* New York: Routledge.
Butcher, Melissa, and Mandy Thomas, eds. 2003. *Ingenious: Emerging youth cultures in urban Australia.* Melbourne: Pluto Press.
Butler, Judith. 1990. *Gender trouble: Feminism and the subversion of identity.* New York: Routledge.
Collins, Jock, Greg Noble, Scott Poynting, and Paul Tabar. 2000. *Kebabs, kids, cops and crime: Youth, ethnicity, and crime.* Sydney: Pluto Press.
Connell, Raewyn. 2002. *Gender.* Cambridge: Polity Press.
———. 2005. *Masculinities.* 2nd ed. Sydney: Allen and Unwin.
Couch, Jen. 2007. Mind the gap: Considering the participation of refugee young people. *Youth Studies Australia* 26(4):37–44.
Crawford, Kate. 2009. These foolish things: On intimacy and insignificance in mobile media. In *Mobile technologies: From telecommunications to media,* ed. Gerard Goggin and Larissa Hjorth, 252–265. New York: Routledge.
Crawford, Kate, and Gerard Goggin. Forthcoming. General disconnections: Youth culture and mobile media. In *Mobile communication: Bringing us together or tearing us apart?* ed. Rich Ling and Scott Campbell. The Mobile Communication Research Series 2. New Brunswick: Transaction.
Cunneen, Chris. 2007. Riot, resistance and moral panic: Demonising the colonial other. In *Outrageous! Moral panics in Australia,* ed. Scott Poynting and George Morgan, 20–29. Hobart: ACYS Publishing.
Evers, Clifton. 2004. Men who surf. *Cultural Studies Review* 10(1):27–41.

————. 2006. Locals only! Presented at the Everyday Multiculturalism Conference of the Centre for Research on Social Inclusion, Macquarie University, ed. Selvaraj Velayutham and Amanda Wise, September 28–29, 2006. http://www.crsi.mq.edu.au/news_and_events/Everyday_Multiculturalism_Conference_Proceedings.htm (accessed September 30, 2011)

————. 2008. Safety on Australian beaches: Terror at Cronulla. *South Atlantic Quarterly* 107(2):411–429.

Fortunati, Leopoldina. 2005. Mobile telephone and the presentation of Self. In *Mobile communications: Renegotiation of the social sphere*, ed. Richard Ling and Per E. Pedersen, 203–218. London: Springer.

Gibbs, Anna. 2001. Contagious feelings: Pauline Hanson and the epidemiology of affect. *Australian Humanities Review* (December). http://www.australian-humanitiesreview.org/archive/Issue-December-2001/gibbs.html (accessed September 30, 2011).

Glazebrook, Diana. 2004. Becoming mobile after detention. *Social Analysis* 48(3):40–58.

Goggin, Gerard. 2008a. Cultural studies of mobile communication. In *Handbook of mobile communication studies*, ed. James E. Katz, 353–366. Cambridge, MA: MIT Press.

————. 2008b. Reorienting the mobile: Australasian imaginaries. *The Information Society* 24(3):171–181.

————. 2010. Official and unofficial mobile media in Australia: Youth, panics, innovation. In *Youth, society and mobile media in Asia*, ed. Stephanie Hemelryk Donald, Theresa Anderson, and Damian Spry, 121–134. London and New York: Routledge.

Gow, Greg. 2002. *The Oromo in exile: From the Horn of Africa to the suburbs of Australia*. Melbourne: Melbourne University Press.

Hjorth, Larissa. 2009. *Mobile media in the Asia Pacific: Gender and the art of being mobile*. London: Routledge.

Höflich, Joachim R., and Maren Hartmann, eds. 2006. *Mobile communication in everyday life: Ethnographic views, observations and reflections*. Berlin, Germany: Frank and Timme.

Ibrahim, Awad. 2008. The new flaneur: Subaltern cultural studies, African youth in Canada and the semiology of in-betweenness. *Cultural Studies* 22(2):234–253.

Jupp, James. 2007. *From white Australia to Woomera: The story of Australian immigration*. 2nd ed. Cambridge: Cambridge University Press.

Kerbaj, Richard. 2006. Warning on African refugee gangs. *The Australian*, December 26. http://www.theaustralian.news.com.au/story/0,20867,20974393–2702,00.html (accessed September 30, 2011).

Kirk, Jackie, and Elizabeth Cassity. 2007. Minimum standards for quality education for refugee youth. *Youth Studies Australia* 26(1):50–56.

Lasen, Amparo. 2005. History repeating? A comparison of the launch and uses of fixed and mobile phones. In *Mobile world: Past, present and future*, ed. Lynne Hamill and Amparo Lasen, 29–60. London: Springer.

Leach, Michael, and Fethi Mansouri. 2003. "Strange words": Refugee perspectives on government and media stereotyping. *Overland* 172:19–26.

Leung, Linda. 2007. Mobility and displacement: Refugees' mobile media practices in immigration detention. *M/C Journal* 10(1). http://journal.media-culture.org.au/0703/10-leung.php (accessed September 30, 2011).

Lilleaas, Ullrich. 2007. Masculinities, sport and emotions. *Men and Masculinities* 10(1):39–53.

Marr, David, and Marion Wilkinson. 2003. *Dark victory*. Sydney: Allen and Unwin.

Olliff, Louise. 2008. Playing for the future: The role of sport and recreation in supporting refugee young people to "settle well" in Australia. *Youth Studies Australia* 27(1):52–60.

Pertierra, Raul. 2005. Mobile phones, identity, and discursive intimacy. *Human Technology* 1(1):23–44.

Poynting, Scott, and George Morgan, eds. 2007. *Outrageous! Moral panics in Australia.* Hobart: ACYS Publishing.

Probyn, Elspeth. 2005. *Blush: Faces of shame.* Minneapolis: University of Minnesota Press.

Prøitz, Lin. 2005. Cute boys or game boys? The embodiment of femininity and masculinity in young Norwegians' text message love-projects. *Fibreculture* 6. http://journal.fibreculture.org/issue6/issue6_proitz.html (accessed September 30, 2011).

Sedgwick, Eve Kosofsky. 1995. *Shame and its sisters: A Silvan Tomkins reader.* Durham, NC: Duke University Press.

Seidler, Victor. 2007. Masculinities, bodies, and emotional life. *Men and Masculinities* 10(1):9–21.

Sheller, Mimi, and John Urry. 2006. The new mobilities paradigm. *Environment and Planning A* 38:207–226.

Thorell, Elin. 2007. *Attending to the needs of refugee youth: The development of a national comprehensive youth service.* Centre for Refugee Research, University of New South Wales.

Tomkins, Silvan. 1962–1992. *Affect, imagery, consciousness.* 4 vols. New York: Springer.

Vincent, Jane. 2005. Emotional attachment to mobile phones: An extraordinary relationship. In *Mobile world: Past, present and future,* ed. Lynne Hamill and Amparo Lasen. London: Springer.

Vincent, Jane, and Leopodina Fortunati, eds. 2009. *Electronic emotion: The mediation of emotion via information and communications technology.* Oxford: Peter Lang.

Westoby, Peter, and Ann Ingamells. 2007. Cross-cultural youth work practice. *Youth Studies Australia* 26(3):52–59.

6 Australian Migrant Children

ICT Use and the Construction of Future Lives

Lelia Green and Nahid Kabir

INTRODUCTION

> I don't have a computer at home. I use it at school. (Male respondent, Australian-born of East Asian background, 16 years)

The past half century has seen issues of globalization increasingly taking center stage alongside the burgeoning of technologies that make communication at a distance more possible. Bryan comments in 1994, before the widespread take-up of the Internet, that it is "the nature of technological development within the area of communications [that] has made physically possible an increasing variety of forms of cross-national communication" (147). The burgeoning possibilities for "staying in touch" with friends and family overseas have added to the complexity of personal relationships for migrants making a home in a new country, as Baldassar has noted (2007). She comments upon a "sense of obligation to remain connected," adding that "the limited capacity for transnational communication in the [immediate] decades following the [second world] war meant that people were not able to, nor obliged to, maintain the level of contact that characterizes transnational family relations today" (Baldassar 2007, 406).

In this chapter we discuss how ICTs influence migrant youth and how these young people negotiate their daily lives through the use of ICTs. First, we discuss the research methodology. Second, we examine Australia as a migrant country. Third, we analyze interviewees' experiences with ICTs. Fourth, we observe how diasporic Australian communities react to world events. Finally, we conclude that many migrant youth negotiate complex identity issues directly or indirectly through their use of ICTs. Given how important it is that young people develop a robust sense of belonging, more research in this area is vital.

RESEARCH METHODOLOGY

The study reported here is a reworking of research originally carried out with a range of thirty-seven culturally and linguistically diverse older children, aged 16–18, still in the school system and living in either Sydney or Perth, Australia. The shared purpose of the two original research projects was to interrogate the sense of identity and hope of Australian children who come from a variety of backgrounds, including some from visible-minority ethnicities and some who adopt markers of religious difference (for example, Australian Muslim girls who wear the *hijab*). The ARC Linkage-funded project "Youth 'at-risk' and 'not at-risk': A study of hope" (Kabir and Balnaves 2005) was conducted in two schools in Perth and was focused on any migrant student irrespective of their religious affiliation. A subsequent Edith Cowan University (ECU)-funded project, "Australian Muslim students: A study of identity, culture and hope," was exclusively conducted with students who identify themselves as Muslim, and involved students from an Islamic school in Perth and eight schools (one faith/Islamic school and seven state schools) in Sydney. For this study we have drawn on respondents from two state schools in Perth and two state schools in Sydney, selecting students who were themselves recent migrants or who had a migrant background.

Both the projects asked similar questions about the students' life stories. The questions were semi-structured, and the discussions open-ended. The questions focused on early childhood memories, family life, parents' work status, students' part-time work, sporting activities, music, interest in computer games, entertainment and cultural interests, religious practices and respondents' hopes, ambitions and dreams. However, the ECU-funded project on Muslim students included some additional questions, such as how respondents defined their identity and sense of belonging, their views on the media, whether the 9/11 tragedy had impacted upon them and how they defined "an Australian" and "un-Australian".

This study was conducted using qualitative research approaches (Strauss and Corbin 1990; Charmaz 2003, 2006) and drew upon the life stories of students aged 16–18 years from migrant families. There is no stronger, clearer statement of how the person sees and understands his or her own life than his or her own narrative of it. The researcher's job, as far as interpretation is concerned, may simply be to identify the meaning or understanding that has already been placed within the story by the teller. Sometimes the stories themselves are so powerful that they need no further interpretation, and the speaker's words provide sufficient insight into how that person understands their life (Atkinson 1998, 64–66). For example, when an "at-risk" youth is asked to indicate the happiest and saddest moment of her life,[1] she might well go into detail about the happy moment but say that her parents' divorce was the saddest moment without elaborating further.

The researcher plays an important role in the process of elucidating participants' life stories (Charmaz 2003, 2006). The question is presented in such a manner that it assists the participant to understand his story, feel the relevance of his life experiences, develop his self-esteem and build up his hope. For example, a refugee from a war-torn zone of Africa trying to settle in a new home in Australia would see the value of his life story when speaking to a researcher who is interested in learning more about him. Supplementary prompt questions are also used. When the researcher asks participants what they do in their leisure time and respondents reply that they watch television or chat with friends through the Internet, then the researcher proceeds further with questions as to whether they watch satellite television, or if they found the computer helped them feel connected to their community and so on. This process of intervention helps develop the interviewee's identity, confidence and hope. As the questions were semi-structured and open-ended, the students felt at ease in sharing their story and directing the conversation. Through answering a diverse range of questions, all the respondents (migrant students "at-risk" and "not at-risk"; Muslim and non-Muslim) revealed that they had adopted a bicultural life style. By biculturalism we mean that they appreciated Western culture such as English language, music, reading, drama and debate, at the same time as proudly retaining their ethnicity and "home" culture. Their bicultural skills helped make them hopeful for their future.

The interviews were tape-recorded and conducted by Kabir, herself from a non-English speaking background. (Kabir was born in Bangladesh and speaks Bengali and Urdu. She lived in Pakistan—then West Pakistan—and the Middle East for several years. So it was easy for her to communicate with students of diverse backgrounds.) Interviews and parental consent were generally facilitated by the schools. Most interviews lasted 30–40 minutes. In this chapter, we focus on the interviewees' responses to questions about entertainment which included discussion of television, movies, computer use and sports. We note that when it came to discussions of ICTs, most students understood these as important in people's everyday life. In this paper we draw upon thirty-seven interviewees' (nineteen students in Perth and eighteen students in Sydney, aged 16–18) use of ICTs. All thirty-seven interviewees were second generation migrants either born overseas or in Australia. The respondents' parents (either one or both overseas-born) are considered to be the first generation migrants to Australia (Kabir and Rickards 2006; Kabir 2007).

DISCUSSION

In Australia, as in many other countries, lack of fluency in the national language confers a range of disadvantages. Further, a number of the children interviewed were from refugee backgrounds and their families had arrived in Australia with few, or no, financial resources. Consequently many of the children included in this study are financially disadvantaged and at the time of the research (2005–2006) were disproportionately less likely to have a computer at home than was the case with mainstream Australian families. In 2007, 91% of Australian families with children aged 8–17 had an Internet connection and three-quarters of these had a broadband service (Australian Communications and Media Authority 2007a, 1–2). This compared with 79% of all Australian households having the Internet at that point, although the proportion of all households which had a broadband service as opposed to dial-up was broadly similar to that of families with children (Australian Communications and Media Authority 2007b, 7). Of the nineteen children interviewed in Perth, four had no computer at home, three had a computer but no Internet access and twelve had the Internet. The eighteen Sydney interviewees included fourteen with the Internet (at least one of whom had dial-up and went to his aunt's home to use broadband), one with a computer and possibly the Internet and three who did not talk about computers or the Internet. In some families the decision not to have a computer may indicate a religious or cultural concern about "content, conduct and contact risks" (Haddon and Stald 2008).

AUSTRALIA AS A MIGRANT COUNTRY

> I have learnt to operate a computer in Kenya, where we lived before coming to Australia. (Female respondent, Sudanese-born refugee migrant, 16–18 years)

With the exception of Australia's indigenous population, who have lived in the country for up to sixty thousand years, Australia is a settler society with a significant proportion of its citizens born overseas. The most recent Census in 2006 (Australian Bureau of Statistics 2007), indicates that 22% of the resident population were overseas-born whereas 44% were either born overseas or had one parent who was born overseas (Australian Broadcasting Company 2009).

For the past decade the country from which the most migrants have come has remained the U.K. (19%) followed by New Zealand (9%), whereas China (5%) has recently overtaken Italy (5%) as the third most numerous contributor of overseas-born Australians (Australian Bureau of Statistics 2007). Although small, with total populations numbering between only 15,500 and 33,000 in a nation of just under 20 million people (Australian Bureau of Statistics 2007a), Australian residents born in Sudan, Zimbabwe, Afghanistan and Iraq are among the fast-growing groups, with significant increases from 2001 in Sudanese- (73%), Zimbabwean- (48%), Afghan- (45%) and Iraqi- (34%) born populations (Australian Bureau of Statistics 2007).

Given the proportions of first and second generation Australians, there has been a significant body of research on the communications patterns of, and connections between, Australian immigrants and their home communities. Baldassar (2007) argues that the five kinds of support identified by Finch (1989) as occurring in proximate (co-located) families are also offered in close-knit, transnational families. These supports comprise "financial, practical, personal (hands-on), accommodation, and emotional or moral support" (Baldassar 2007, 389). They flow both ways depending upon the circumstances and are subject to continuous negotiation where good communications channels exist. Because of the distance barriers experienced by transnational families, especially where members live in an isolated island continent like Australia, Baldassar identifies "communication technologies" and "travel technologies" as being crucial to delivering and receiving these supports in full. The communication and travel technologies enable, between them, two forms of contact: "virtual contact" and "face-to-face contact".

Whereas travel technologies are restricted to those wealthier and more settled migrants, virtual contact is an important element of caregiving and receiving between families connected across national borders. According to Baldassar (2007, 391–394), three types of care are exchanged between adult children and their transnational kin located outside Australia: "routine" care, as in regular calls or e-mails which "keep in touch"; "ritual" care, communications which mark birthdays, anniversaries, saints' days and so forth; and "crisis" care, "unexpected or unanticipated events or times of increased need," such as an acute illness, or at the time of a relationship break-down such as a divorce. For some migrants in Australia, well aware of their relative wealth, the obligation to send financial support back to relatives living in their home country is keenly felt. Remittances, which flow from wealthier countries to poorer countries, are estimated to help support 10% of the world's population (International Fund for Agricultural Development 2007; Green 2010, 173–175) and have been called "the human face of globalization". They have been described as the only economic indicator that rises in times of global financial turmoil. Wilding (2006) notes how an Australian Iranian refugee married to an immigrant from a wealthy Western country had threatened to leave her husband if he kept sending money to his comparatively wealthy family overseas, "saying that they did not have enough money to provide for their own children or their family in Iran, let alone money to send to extended kin living in relative safety elsewhere" (135). The communication patterns of minor-aged migrant children are under-researched, but it is against this background of family interactions that any examination of children's ICT use must be positioned.

INTERVIEWEES' EXPERIENCES WITH ICTS

We have a satellite [TV] at home. So I watch the Arabic shows because I don't want to miss the shows that I used to watch there. Every night I watch a serial if I can. (Female respondent, Middle Eastern-born refugee student, 16–18 years)

The children interviewed in this study divide into two major groups around the axis of personal cultural experience. If they were old enough to be well-integrated within a peer network in their country of origin, or their original place of departure for Australia, then there are indications that they use ICTs to try to retain some links with that cultural context. In this sense, they may be re-enacting as a younger generation the interconnecting with social networks elsewhere which is carried out by their parents and which has been documented by Baldassar (2007), Wilding (2006) and others. Indeed, Wilding cites Glick Schiller, Basch and Blanc-Szanton (1992, ix) to talk about how migrants use "ICTs to create, support and reproduce 'social fields that cross geographic, cultural and political borders'" (Wilding 2006, 137). Although there has been little work on the efforts of minors to reproduce these social fields, the indicators that they do so are clear.

A Hong Kong-born female student talks about ICTs: "I have a computer at home. I chat with my friends in China through the Internet. Sometimes I play computer games." Asked about her TV viewing habits she comments that, as well as watching Australian programs in English, she watches "Chinese programs through the satellite TV." There may be a second reason why this interviewee seeks connection through her ICT use: her mother has remarried and she does not get on with her step-father, who is (she says) "a rude person. Under him we live in a state of terror." Apparently, as well as experiencing some cultural dislocation in terms of wishing to remain in touch with friends in Hong Kong, this teenager is experiencing the frustration with adult carers common to teenage experience the world over—assuming there is no more justifiable reason as to why she has such a negative view of her step-father.

Similarly, a 16–18-year-old Vietnamese-born male migrant, whose family entered under the Business Migration scheme, says that he uses ICTs to manage the distance between him and his friends: "I miss my friends in Vietnam. I chat with them through the computer, so it is not so bad. I watch news sometime through web page." This account of an interviewee "missing" his former friends contrasts with an Australian-born young man, with one parent born in Malaysia: "I have a computer at home. I play some games. I have Technical Graphics in my TEE [final school, university entrance exam] subjects." Asked about television viewing, this interviewee says, "Yeah, I watch, but I don't really have a favorite TV show;" when asked about his hobbie, he says: "I love to draw sketches in my spare time and play other sports." In this case, the list of preferred activities does not clearly refer to the development and maintenance of a Malaysian cultural heritage but might instead be interchangeable with the activities of most other adolescents of an equivalent age.

A Sydney-based Australian-born male migrant of Mauritian background says, "We are into Mauritius culture." This is communicated as a family choice, however, rather than an expression of the young person's developing sense of self. Instead, the issues around this adolescent's ICT use are rather more universal and prosaic: "I have a computer at home. I don't play games on the computer. I used to play a lot when I was in year 8 but I think my parents got a bit strict on me; then I had a lot of work to do. Then eventually, I found other things to do besides playing games." Asked about leisure time he says, "If I come back from school, mainly I listen to music, go on the Internet, work out, do my homework, talk to friends [and] go out."

This list of teenage activities contrasts with another Australian-born interviewee of Middle Eastern background, whose parents are both strict Orthodox Christians and who is having issues with them about family restrictions, such as not being allowed a boyfriend until she is 18. This respondent did not have access to the Internet at home:

> I am not too much of a computer person, but I like computers for researching and doing my work. We have one at home but I never use it. I just need someone to basically go through the computer stuff with me. We don't have the Internet at home; I use the Internet at school. I watch TV at night time but mostly do my homework.

The implication here is that the respondent sees little point in using the computer given that it is not connected to the Internet.

Other school students who effectively had little computer access included an Ethiopian-born refugee girl who comments, "In my house I do not have a computer [although] I know how to turn on the computer. I cannot type fast because I don't have a computer." A Middle Eastern-born refugee boy whose capacity to use the computer is restricted as a result of his parents' communications needs says, "We have a computer at home but the keyboard is in Arabic. I can't write letters in Arabic, so I don't use it. I use it at school. On the computer, I only play card games." A Rwandan-born female refugee also comments, "I do not have a computer [although] I like computers. It is good for me. I am trying to learn how to work on it." In terms of media choices she notes, "I watch *Everybody Loves Raymond* and *Simpsons*, and we hire African [Nigerian] movies from the video shops."

An overseas-born refugee girl from the Balkan states said, "Yeah we do have Internet but . . . It's expensive though. Yeah, we buy like a card $10—we don't have connection or anything." This respondent, and her older brother, have part-time jobs which provide the family's major source of earned income. Another interviewee, an Australian-born student of Lebanese background who has a dial-up Internet connection, commented that he had "got the Dodo one, but it comes every month I think; yeah something like that. But we hardly go on it because my aunt's got the Internet. And she's got a broadband and we just go to her house every now and then." In this family, it is the satellite TV subscription that takes financial precedence over the possibility of a broadband connection. The young man went on to say, "Sometimes I just watch the Arabic channels and sometimes I watch 9, 10 and Fox channels. Movies and. . . and when we are bored and have nothing to do then we watch the dish . . . I listen to Arabic music." A young woman of similar heritage, also born in Australia, nonetheless suggested that a shared cultural background was an important unifying factor for her local friendship group: "I like computer games. Any games like sport games, stuff like that. Or just I go on MSN . . . chat with friends. [My] friends are all from an Arabic background; we speak a bit of Arabic." For some of the interviewees, the computer is a tool for creativity, like the young Australian-born man with a Malaysian background who uses his computer to further his studies in Technical Graphics, and an Australian-born woman with a Lebanese background who talked about the use of her computer for Photography: "I have Photography as a subject. [We work] just like artists, and just doing digital work . . . It's just like what you can do with images, and it's basically art too; it's the same thing. How you can change the art and make something unusual and things like that." An Australian-born female student of Bangladeshi background also commented, "I have Digital Arts as a subject," indicating that this was a major reason for the home-based computer. An overseas-born male student of South American background also commented, "Tech Graphics is my TEE subject. I like Tech Graphics because it gives me a chance to design and use my imagination. I think I am going to be a product designer." Similarly, an Indian-born female migrant saw her future in digital culture: "Yeah I'd like to be a multimedia developer." Another Australian-born female student of Southeast Asian background said:

> For my subjects I chose Photography and Computer Studies. Um, it's to do with, the thing it's taught me, is that I love about it and you learn stuff on computers; how you manipulate things, and yeah. It is normally [used for] computer designing. Photography—through the digital things you do. Yeah and we also make animations through it. I made an artwork. In the artwork I tried to show that the artwork has power.

Sometimes the students' preferences, and their common ground with their peers, had been challenged by the parents' ICT choices: "Yeah I'm a Bollywood sort of person, but . . . English TV I can't watch . . . I used to watch *Neighbours* a lot, but not any more, because my mum put the dish on! So she watches that, so now I watch all the Indian dramas."

In these vignettes, both overseas-born and Australian-born migrant children demonstrated many of the themes universal to the teenage experience. Regardless of gender, they used their passions and enthusiasms to indicate possible future careers; they realized their parents still had a huge say in what they do and what their priorities should be; and some chafed under parental restrictions. In other ways, however, these children had a very different experience from their mainstream Australian peers. Their financial circumstances often precluded access to the range of ICTs present in wealthier homes. Many migrants are economically disadvantaged compared with mainstream Australian citizens. For example, in 2001 and

2006, the unemployment level for Muslims in Australia, many of whom are first or second generation migrants, was three times higher than the national average (Kabir 2008, 400).

Even where there was the financial possibility of subscribing to broadband Internet, this was sometimes decided against in favor of a satellite TV subscription that would allow one or both parents to feel more connected with their home culture and community. For some respondents, especially those who were older when they first migrated to Australia, the Internet provided a lifeline to friends from their previous life. Clearly, this is not an exhaustive study but it is indicative of the range of issues that is likely to arise in a dedicated investigation of migrant children's ICT use.

DIASPORIC COMMUNITIES

> The thing is that we watch the Indian dish antenna. We watch that.
> (Female respondent, Iranian-born, 16–18 years)

Many of the students interviewed were using ICTs to negotiate a dual identity: Australian on the one hand and membership of a diasporic community on the other. Wilding's (2006) research with adult migrant children and their older parents overseas identified that a wide variety of technologies were used to stay in touch with extended family members:

> Interviewees talked of using a wide variety of methods in the past and present, including (but not limited to) telephone calls, telegrams, letters, faxes, email, Internet chat rooms, Internet websites, mobile text messages, videos, tapes, gifts, cards and postcards. (129–130)

Through exposure to these communication flows, and through their consumption of satellite television programming that taps into a non-Australian cultural background, some of these young people were increasing the common bonds with their countries of origin rather than with their Australian contemporaries. In doing this, these migrant children were consuming, and contributing to, what Dayan (1998) in his essay "Particularistic media and diasporic communications" has termed a "micro public sphere" (103).

In Dayan's terms, prior to the widespread adoption of the Internet, the particularistic media that constitute the micro public sphere include newsletters, audio and videocassettes, holy icons, photographs, telephone calls, letters and exchange of travelers. The aim of these particularistic media is to circulate and cement understandings between people who already acknowledge a shared past. It is not to bring a new community into being, but is instead an attempt to stop an extant group from disintegrating. These are "media whose aim is not to create new identities but to prevent the death of existing ones" (Dayan 1998, 110). As indicated in the range of responses offered by the young people in this study, and as may be indicated by possible differences between overseas-born and Australian-born migrants, association with a diasporic identity is to some extent a matter of individual choice as well as personal circumstance. Dayan's view is that "a diaspora is always an intellectual construction tied to a given narrative" (Dayan 1998, 110). This indicates that a willingness to own one's part in such a narrative is an important element in the adoption of a diasporic identity. At the same time that these migrant children are learning about their (new) homeland of Australia, most are also learning how to be members of a diasporic community within Australia and how to use ICTs to that end.

For example, many migrants keep themselves updated on world events through the use of the Internet and through satellite television channels. Such connections mean that young migrants, as well as their parents, can get distressed when an upheaval impacts on family members living overseas. For example, during the 2006 Lebanon War, a very distressed Australian-born student of Lebanese background said:

> My grandma's there [Lebanon], like as I told you, my mum's whole generation is there, but we try to call them and there's no reception. It's really hard for them to speak, but you can, like my mum tries to speak to them and it's really hard for her to speak because she can't understand them. And like she keeps trying to call, keeps trying to call, keeps trying to call and it keeps hanging up. Like last night when she tried to call, they answered and she spoke to them really good. And she just asked them "How've you been," then she goes "We can't walk outside, we can't do a thing." (Cited in Kabir 2008, 416)

New media allow the emotional involvement of the diasporic community in distant events and allow the younger generation to voice their opinions on contemporary issues.

Dayan recognizes the continuous negotiation that these different pressures require in terms of the complexities of potential tensions between an adopted identity and an inherited cultural identification. He addresses the "production", "confrontation" and "adoption" of identity, taking into account the capacity of the host community to make assumptions based upon cultural and linguistic differences and "assign" an identity. This may be accepted as unproblematic by the assignee, or it may be confronted and resisted. Resistance acknowledges that the identity forged within a diasporic community is rarely entirely in accordance with that assigned by the majority culture. The "owning" of an individual identity involves both an awareness and response to the public sphere of the wider host culture and an engagement with the private sphere of the diasporic person's life; it is "where the intimate meets the historical." Dayan (1998, 107) refers to the process of owning a diasporic identity as the "rediscovery" or "reinvention" of tradition and it may be represented by individual choices about dress codes, language use, dietary habits and religious identification. Older migrant children, such as those considered here, are in the process of making such decisions for themselves using the ICTs at their disposal both at home and at school.

CONCLUSION

Although this is still an under-researched topic, it is to be expected that studies of migrant children's use of ICTs will reflect the particular circumstances of the child as a young person growing up in a family, and struggling to assert their independence. This theme, which is common to most older adolescents, is nonetheless played out against a background of individual difference which recognizes the specifics of cultural heritage and the pressures imposed by the country of adoption. In the negotiation of identity, the age of the child at the point at which they enter their new country of adoption is likely to be a critical factor, as are the comparative resources available to the family to fund ICT use.

Migrant children usually grow up in migrant families where parents or caregivers are actively involved in negotiating the absence of extended family. Through watching their parents' communication patterns, and through living with their parents' priorities in terms of media and technology choices, migrant children learn what it is to express cultural difference and a diasporic engagement within the context of the majority culture. For a portion of young people, ICTs can become a means of escaping the restrictions of the family and engaging with friends from the mainstream culture outside the diasporic context. For some, their ICT use allows them to seek refuge in the familiarity of the old language and friendship system, whereas for others it provides an opportunity to keep abreast of world affairs as a means of exercising their rights as global citizens with opinions on political activity at home and abroad.

The complex balancing act implicit in these different uses of ICTs and different responses to the host community and to the culture and community of origin deserve further consideration and investigation. Even so, the indicators are that ICT use expresses the individual as they are, even as it helps to create the future for that individual as they will become.

NOTES

1. The pronouns "she" and "he" are both used in this paper and, where appropriate, should be taken as including members of the other gender.

REFERENCES

Australian Broadcasting Corporation (ABC). 2009. Census figures reveal quarter of population born overseas. *ABC News*, January 29. http://www.abc.net.au/news/stories/2009/01/29/2477425.htm (accessed April 2009).

Australian Bureau of Statistics (ABS). 2007. More than one in five Australians born overseas: 2006 Census. ABS Media Fact Sheet, June 27. http://www.abs.gov.au/AUSSTATS/abs@.nsf/7d12b0f6763c78caca257061001cc588/ec871bf375f2035dca257306000d5422!OpenDocument (accessed April 2009).

———. 2007a. 2006 Census QuickStats: Australia, released 27 June 2007, http://www.censusdata.abs.gov.au/ABSNavigation/prenav/ViewData?subaction=-1&producttype=QuickStats&areacode=0&action=401&collection=Census&textversion=false&breadcrumb=PL&period=2006&javascript=true&-navmapdisplayed=true& (accessed September 2011).

Australian Communications and Media Authority (ACMA). 2007a. Media and communication in Australian families 2007. http://www.acma.gov.au/webwr/_assets/main/lib101058/media_and_society_report_2007.pdf (accessed September 2011).

———. 2007b. Telecommunications today: Consumer attitudes to take-up and use. http://www.acma.gov.au/webwr/_assets/main/lib310210/telecomms_today_consumer_takeup_and_use_of_tcomm_svces.pdf (accessed April 2009).

Atkinson, Robert. 1998. *The life story interview.* Qualitative Research Methods Series 44. London: Sage Publications.

Baldassar, Loretta. 2007. Transnational families and the provision of moral and social support: The relationship between truth and distance. *Identities: Global Studies in Culture and Power* 14(4):385–409.

Bryan, Dick. 1994. "The multilocals": Transnationals and communications technology. In *Framing technology: Society, choice and change*, ed. Lelia Green and Roger Guinery, 145–160. Sydney: Allen and Unwin.

Charmaz, Kathy. 2003. Grounded theory. In *Qualitative psychology: A practical guide to research methods*, ed. Jonathan Smith, 81–110. Thousand Oaks, CA: Sage.

———. 2006. *Constructing grounded theory: A practical guide through qualitative analysis.* London: Sage publications.

Dayan, Daniel. 1998. Particularistic media and diasporic communications. In *Media, ritual and identity*, ed. Tamar Liebes, James Curran, and Elihu Katz, 103–113. London: Routledge.

Finch, Janet. 1989. *Family obligations and social change.* Cambridge: Polity Press.

Glick Schiller, Nina, Linda Basch, and Christina Blanc-Szanton. 1992. Towards a definition of transnationalism. In *Towards a transnational perspective on migration: Race, class, ethnicity and nationalism reconsidered*, ed. Linda Basch, Nina Glick Schiller, and Christina Blanc-Szanton, ix–xix. New York: New York Academy of Sciences.

Green, Lelia. 2010. *The Internet: An introduction to new media.* Oxford and New York: Berg.

Haddon, Leslie, and Gitte Stald. 2008. A comparative analysis of European media coverage of children and the Internet. Paper presented at the Association of Internet Researchers (AoIR) Annual Conference, October 16–19, Denmark, Copenhagen. http://www.lse.ac.uk/collections/EUKidsOnline/EU%20Kids%20I/Presentations/EUKidsPressAnalysisAoIR.pdf (accessed September 2011).

International Fund for Agricultural Development. Sending money home: Worldwide remittance flows to developing countries, 2006. 2007 International Forum on Remittances. http://www.ifad.org/events/remittances/maps/ (accessed April 2009).

Kabir, Nahid. 2007. What does it mean to be unAustralian? Views of Australian Muslim students in 2006. *People and Place* 15(1):51–68.

———. 2008. To be or not to be an Australian: Focus on Muslim youth. *National Identities* 10(4):399–419.

Kabir, Nahid, and Mark Balnaves. 2005. Students "at risk": Dilemmas of collaboration. *M/C Journal* 9(2). http://journal.media-culture.org.au/0605/04-kabir-balnaves.php (accessed September 2011).

Kabir, Nahid, and Tony Rickards. 2006. Students "at-risk": Can connections make a difference? *Youth Studies Australia* 25(4):17–24.

Strauss, Anselm, and Juliet Corbin. 1990. *Basics of qualitative research: Grounded theory procedures and techniques.* Newbury Park, CA: Sage Publications.

Wilding, Raelene. 2006. "Virtual" intimacies? Families communicating across transnational contexts. *Global Networks* 6(2):125–142.

Theme 3

Looking at the Migrations and Diasporas Through the Lens of the New Media

7 Diasporas, the New Media and the Globalized Homeland

Raul Pertierra

INTRODUCTION

This paper investigates the consequences of the out-migration of large numbers of Filipinos (25% of the working population) in search of work, to escape political oppression, to seek better lives or simply out of a sense of adventure and desire to explore the world. These factors are indissolubly linked and their corresponding motivations and identities inseparable. The communications revolution has only strengthened these parallel forces of attraction and repulsion. For this reason, migration in all its forms, technological change, global security and sustainability should be investigated as an interacting system whose parts complement as well as disaggregate their respective elements. People have multiple identities and develop them according to specific needs and contexts. It is the socio-economic structures themselves that oblige people to invent themselves according to particular requirements.

In this framework the paper discusses some of the ways in which the new global condition affects notions of identity and cosmopolitanism. My approach employs the notion of concrete theorization—one where theoretical questions are embedded in concrete and particular contexts (Tomlinson 1999). Rather than provide specific policy suggestions for the problems of migration, refugees and globalization, this study examines the conditions underlying contemporary transhumance and their implications for notions of sustainability. It is for this reason that the paper combines elements of both topics—migration and technological change. The approach is basically sociological, with a strong emphasis on close ethnographic description and the contextualization of aspects of everyday life. Globalization has to be understood both at an international level and as a local condition. While the former approach requires structural models, the latter is best understood ethnographically. The chapter explains and then illustrates the new media with examination of social and cultural practices of Filipino citizens that extend internationally through the diasporic communities.

THE NEW MEDIA

Although the application of information technology (e.g. Internet, mobile phones, etc.) is often similar, their specific social and cultural impact varies locally, as these technologies are influenced by particular cultural traditions, power structures and economic resources. Societies and information technologies engage dialogically, where each is shaped by the needs of the other. This, however, does not mean that societies will be determined by information technologies or that societies can dictate the course of ICT development. The case of mobile phones in the Philippines demonstrates how a globally introduced technology interacts with local socio-cultural conditions. This interaction is not only materialized in indigenized applications of the global technology (e.g. txt God) but also expresses how modernity, as a social condition of globalization, is manifested in a Philippine context. Technological modernity in Asia is not always accompanied by Western secular structures and beliefs. Instead, the supernatural is adapted to the new technology. The mobile is simply another medium for manifesting supernatural powers.

The concept of the digital divide refers to inequalities in the access and use of CMICTs (Computer-mediated interactive communication technologies). The digital divide or digital inequality is not only about the existence or non-existence of infrastructure. The provision of physical access is necessary but it is not sufficient. Digital inequality refers to "real access", which includes cognitive and cultural capital in addition to technical resources. Real access goes beyond infrastructure and refers to people's actual possibilities to use technology to improve their lives. The technology itself does not ensure its equal and efficient use but real access is ensured only when appropriate technologies are introduced in political, economic and social environments conducive to people's usage (Kuvaja and Mursu 2003).

Mobile phones are increasingly taking a major role in this communication revolution and are often referred to as the "new media" (Computer-mediated interactive communication technologies, CMICTs). The interconnectivities of mobile telephony, desktops, radio, television and print are producing new communication structures with often unpredictable consequences (e.g. smart mobs, prosumers, citizen-journalists, new media stars). But it is certain that the use of the new media will bring about important social changes at distinct levels of social structures, from personal identity to political mobilization, from consumer choices to global corporations, from virtual spaces or simulated models to lived realities and embodied geographies. In the following sections I explore these new media environments and discuss three case studies taken from my research.

COSMOPOLITANISM AND GLOCALIZATION

This new informational order is usually seen as a process of cosmopolitanization, but it is also associated with the return of locality and the revitalization of diasporal and virtual communities. Until recently, the new media mainly involved the Internet and its associated technologies. The term invoked cyberspace, constructed as a virtual environment detached from the physical world, a place where people could inhabit and create new types of communities. Often this involved hybrid spaces combining the global with the local ("glocal").

Virtual communities have often been studied as narrative spaces where users create collective environments composed mainly of texts. More recently, the idea of a virtual world, a simulated space, completely disconnected from our physical environment is challenged by the emergence of mobile devices. These communities of phoneurs result in mobile collectivities whose member inhabit embodied but transhumant spaces. Unlike desktops and other immobile technologies, mobile phones more closely resemble tools or prosthetic devices as extensions of the body. They become extensions of the hand, allowing us to connect anytime, anywhere with anybody. Bodies themselves become writing devices as phoneurs negotiate new urban spaces (Fortunati, Katz and Riccini 2003).

Urbanity itself becomes a feature of the new mobility. This new urbanity is complemented by other communication technologies such as video, MP3 players and other multimedia interfaces. With the aid of nomadic technologies, virtual social communities emerge in physical spaces. What earlier were post-corporeal, non-spatial communities now emerge in specific localities. In this context, CMICT plays an active role in creating new types of communication and social networks. As Kopomaa (2000) argues, the mobile puts the city in your pocket.

NOMADIC TECHNOLOGIES AND NEW PHYSICALITIES

Cyberspace has frequently been regarded as a utopian space in which users are able to project their imagination and communication. When communities are shaped in a hybrid space, CMICTs become new tools for creating novel and unpredictable imaginary spaces, renarrating lived space (Kirby 1997). While fixed Internet users do not have the ability to move through physical space, the emergence of nomadic interfaces makes possible mobile imaginary spaces to be enacted and constructed in physical space. Hence, nomadic technologies have a role in the construction of narrative spaces. They allow virtual spaces to be mobile, bringing them into the physical world.

Location awareness embedded in mobile devices strengthens the connection to physical spaces, creating new geographies of mobility. Mobile devices and interfaces make us aware of the importance of physicality when dealing with digital spaces. It is in this sense that mobile phones can be perceived as writing devices. Writing in a broader sense (not only SMS or MMS) means the creation of narrative and imaginary spaces. Cell phones are new media devices writing in both physical and hybrid spaces transforming them into textographic spaces (Josgrilberg 2008). Rather than converting space into text as did the earlier technology, the new media, including Web 2.0, reintroduce the text into embodied and lived spaces.

Pingol (2006) describes how Filipina women married to Korean men use the mobile to ensure their own safety. Using their mobiles, Filipinas in Korea band together and keep track of their friends' activities:

> A Korean husband begged me to help him after learning that I was looking for Filipinas married to Koreans. He brought out his mobile and gave me his wife's number . . . My attempts to connect with his wife went unheeded. This was not unusual, I was told by friends of battered wives. The woman in hiding makes a point of not responding . . . taking a call is always a risk. She must be with another Filipina willing to provide her a temporary home. (Pingol 2006, 56)

MIGRANTS AND OVERSEAS WORKERS

It is estimated that over 8.5 million Filipinos (12% of the Philippine population) live abroad, either as permanent migrants, overseas workers or refugees. This is a drain on the country's human resources but also the salvation of its local economy. This movement of Filipinos going overseas is caused by several factors. Among them, the deterioration of local living conditions (e.g. poor wages, personal insecurity, political instability) and the overly positive evaluation of life abroad are the most prominent. Both are consequences, although at different levels, of globalization. The problems of everyday life and the fantasies of alternative lifestyles combine to produce an irresistible attraction to seek other places—elsewhere is always better (Pertierra et al. 1992). A leading housing company advertises its products as "making you feel you are living abroad." The difficulties of everyday life and its accompanying imaginaries are resolved through migration or overseas work. Filipino doctors retrain as nurses to improve their employment opportunities abroad. Qualified teachers work as domestic helpers in Hong Kong to send their own children to private schools.

As significant dollar earners, Filipino overseas workers are considered "modern day heroes", a term coined and ascribed to them by the Philippine state. Recently (2008), using a more glamorous term, President Arroyo referred to them as expatriates rather than Overseas Filipino Workers (OFW). However, there is an absence of empirical research looking into the applicability of these state narratives as they affect individual identities. Except for mostly professional workers, many Filipinos experience deprivation and difficulties abroad.

COMPUTER-MEDIATED ICTs AS TECHNOLOGIES OF MEDIATION

Most of our informants welcome the advantages of mobiles and the Internet. While they also recognize some problems such as rising costs (e.g. over P300 monthly for mobiles), the lures of gambling or the dangers of seduction, they overwhelmingly support their advantages. Remaining in contact with friends and kin or extending a social network, accessing useful information, including spiritual sites or playing games are among the most significant uses of the new technology. While mobiles have blended into the routines of everyday life, the Internet remains less accessible because of economic and technical reasons. But Internet access is readily available in most urban centers and young Filipinos quickly learn the skills needed to navigate in the cyber world.

The experiences made possible by mobiles and the Internet are encouraging new forms of individualism and cosmopolitanism. Strangers are increasingly entering networks of intimacy hitherto limited to kin and close friends. These extended networks require individualizing responses and broaden outlooks, leading to more cosmopolitan orientations. Online relationships are the best expression of this new cosmopolitanism. Many Filipino women routinely use the Internet to explore and widen prospective marriage choices (Constable 2005). They also use it to access religious sites, indicating that this technology is also compatible with traditional notions of the authentic.

Intimate Connections

The requirement for privacy is crucial for many aspects of modernity. The modern subject is expected to cultivate interests, needs and desires which mark him/her as a particular individual. In most Asian societies, the requirements for a private individualism have only developed recently. Fashion and the accumulation of material goods have become a major concern for these affluent Asian societies. Consumerism is an ideal activity within which to cultivate personalized identities. In an otherwise mass society, personal patterns of consumption counteract feelings of collective anonymity.

The Philippines has so far not shared a general affluence, and the scope for cultivating a private individualism has been limited. Only members of the middle class have had the luxury of sufficient domestic space and access to the telephone to cultivate a private self. Others, such as those living in rural communities, only experience these possibilities when they obtain overseas work. On their return, the display of private goods and foreign tastes celebrate and mark their new identities (Pingol 2001).

While lacking domestic space to develop private liaisons, many Filipinos have turned to mobile phones as another alternative. Texting has become the major way for most Filipinos to cultivate a network of acquaintances known only to them. Texting most often involves relatives, common friends and associates, but it also includes strangers. The latter opens possibilities for intimate and private identities. While contacting strangers was possible in the past, texting provides the anonymity, privacy and convenience not hitherto available.

The mobile allows Filipinos to cultivate personal and private relationships. Bored housewives, inquisitive teenagers and men seeking sexual satisfaction increasingly use these services to contact like-minded individuals. Mario texts a 41-year-old woman from Laguna, whose husband lives away from home—they discuss their sexual needs casually. Sherry, a young model, and Carlos arrange to meet for a date shortly after exchanging texts with one another. Rems, whose husband is also away, engages in sexual repartee via texts with Joel while accompanying her sister shopping. All these examples indicate the facility with which unsurveilled communications can occur in texting. Dave regularly takes advantage of any opportunities for sexual experiences through texting. He readily accepts mis-sent messages to engage in "text sex" (Pertierra 2006). While all these cases include explicit sexual elements, they are as often ludic as sensual. They are also good examples of technologically mediated relationships. Such mediated relationships increasingly characterize our lives, enriching its possibilities but often substituting the real with its simulacrum.

The Gift and Communitas

An important feature of the new media is the so-called practice of gift exchange. Text messaging in the Philippines and other uses of the new media often rely on freely given information or services. This expresses the personal ties often linking their users. A model of amity is extended even to strangers, where personal information and advice is shared with a network, many of whose members are unknown to one another. This is a form of cyber gossip where sharing information is an instance of group inclusion. Many of these cyber communities, apparently characterized by a free exchange of gifts and services, approach what anthropologists (Turner 1974) call "communitas", whose members momentarily suspend existing hierarchies and divisions, co-mingling freely for the duration of the ritual. Communitas is inevitably short-lived and only serves to reinforce and reinvigorate existing divisions by momentarily suspending them. But the memory of its experience elicits aspirations for an alternative future.

The Internet café in the Philippines is a physical site for communitas. Unlike their Western counterparts, Philippine Internet cafés are sites for both corporeal and virtual consociation. The normal constraints of gender, class and generation are suspended in the café, allowing alternative relationships, both real and virtual, to develop. Like the London coffee houses in the 17th century, Internet cafés facilitate communicative exchanges in the real and virtual worlds.

Janna experiences a new sense of agency when she goes to the Internet café. Her competence in gaming gives her a new confidence and allows her to make hitherto unlikely friends. Married Filipinas generally avoid friendships with men who are not their kin. The Internet café is a place where people from widely different backgrounds develop both real and virtual ties. Even users with online connections at home prefer to use the café for privacy and sociality. The novelty of the Internet café allows gender, class and generational roles to be relatively unaffected by traditional constrains. Genuine, another Internet user in the same café, admits that he prefers his café and online friends over his classmates:

> Yes I feel a sense of community online. Actually you can find more people of the same characteristics in chatrooms than in 'real' life. For me, I felt more loved and more bonded with friends online than offline. And my circle of friends grew. (de Leon 2007, 49)

What these examples show is the development of a local cosmopolitanism based on extensive and variable relationships. Online marriages are the most visible examples of this phenomenon. Constable (2005) discusses how supposedly naïve Asian women often show more sophistication and awareness than their male Western partners. Moreover, contrary to the early depiction of cyberspace as disembodied, Constable stresses the close correlation between actual bodies and their representations on marriage sites. Chen (2004) makes a similar point for Taiwanese men seeking Vietnamese wives online.

GLOBAL LURES AND DISTANT HOMELANDS

The Philippines presents a paradigmatic example of the aporia of globality—the desire, obligation or compulsion to leave one's home community in order to attain a better life, while still remaining attached to one's roots. This attachment is made easier by the facility of remaining in touch. Connections need no longer be severed or even decreased in the age of global communication. The good life is often defined as outside one's community while one's identity remains rooted within it. The new communication media allows one an absent but continued presence. Once abroad, Filipinos complain of home-sickness and exert great efforts to communicate with friends and family. CMICTs provide new channels of communication that, paradoxically, both alleviate and enhance feelings of home-sickness. A major dilemma of globality is that "home" is permanently elsewhere.

In this chapter culture is seen as something made rather than found—no longer spatially fixed—but a social and historical performance (Pertierra

2002). Culture as construction or performance implies that identities may be temporary, fluid and heterodox or hybrid, where the actor is an agent *"evading or manipulating every national, ethnic, or religious identity."* (Gupta and Fergusan 1997, 45) Palestinian refugees imagine the "home" and "homeland" as a romanticized lost lover while the Japanese in California recreate "Asian landscapes" (Gupta and Ferguson 1997). Pertierra (2002) reports similar expressions of nostalgia among Filipinos settled in Toronto whenever they recall village life. They often use the metaphor of being drawn to their villages like fish in search of spawning grounds. They attempt as much as possible to reproduce village life in Toronto. Apart from remaining in close touch with village kin, they also establish neighborhoods consisting of other Ilocanos. They obtain work for one another and engage in common leisure activities. While apparently reproducing traditional cultural forms, Filipinos abroad are aware of the new context within which these forms are enacted. In the case of permanent migrants, the challenge is adjusting to the host culture as well as reconstructing aspects of a previous identity.

Ravindran (2008) has addressed this conundrum by arguing that the globalized homeland now includes its diasporal members. Distinctions between the settled and the original homeland no longer apply as they did before CMICT. A friend describes herself as a Filipino from California, converting what was hitherto a nationality into an ethnicity. This transition is seen in the change from Filipino American (1st generation migrant) to American Filipino (3rd generation). A growing Filipino diaspora will similarly interrogate earlier definitions of homeland. The settled and the original homelands now overlap within a globalized reality.

Flexible Networks Rather Than Solid Structures

This globalized homeland is no longer adequately described as consisting of stable structures but rather of shifting and transforming networks or landscapes (Appadurai 1996). Castells (2001) suggests a new model for these landscapes. Rather than seeing society as arising out of fixed structures shaping identities and generating practices, individual and collective ties are better seen as complex networks linking persons and communities in multiple alliances and shifting formations. Ethnicity was an earlier expression of these shifting identities previously linked to territories but quickly adaptable to transhumant populations. People take their identities with them as they move around but adjust them to prevailing conditions.

Hannerz (1993) describes how culture, hitherto collectively shared, is now individualized. "As she changes jobs, moves between places, and makes her choices in cultural consumption, one human being may turn out to construct a cultural repertoire which in its entirety is like nobody else's" (Hannerz 1993, 105). While the uniqueness of individual experience is a feature of all societies and cultures, the contemporary emphasis on consumer choice and the wide variety of patterns of consumption available make this insight particularly relevant and appropriate for our times.

Globalization and TV Audiences

Remittances from overseas Filipinos now constitute the largest source of overseas income for the Philippines ($14 billion, 2007) far surpassing foreign aid and investment ($650 million, $1.5 billion respectively, 2007) (Philippine Overseas Employment Agency Report 2009, 232). While these remittances undoubtedly contribute to the local economy, their developmental effects appear minimal. Remittances are mostly used for daily expenses or for prestige and status rather than investment. In fact, many of the skills acquired abroad are often lost when workers return to their home communities (Pertierra et al. 1992). But their contribution is much more than the merely economic. Socio-cultural changes are also underway as a result of this massive movement of people abroad. Filipinos are one of the largest Asian migrant groups in the U.S., where they play an important role in healthcare. Filipinos also form substantial communities in Canada, Australia and New Zealand. Their English fluency, relatively good education and their willingness to assimilate are the major factors for the success of Filipino migrants.

In a recent episode of the popular television series *Desperate Housewives*, the character Teri Hatcher was diagnosed as entering menopause. In disbelief she exclaims, "Before we go any further, can I just check those diplomas? I'd just like to make sure they're not from some med school in the Philippines" (June 26, 2009: 12). This brief remark in a fiction television series, a parody of American suburban culture, provoked outrage among many Filipino medical practitioners in the U.S. They claimed that the remark maligned their competence and soiled their reputations. In a disputatious and politically correct society such as the U.S., such a response may not seem unusual but even by local standards demands for compensation ($500 million) appear excessive. As often happens in such cases, the outrage quickly gathered pace among American-based Filipinos and soon enough some Philippine medical institutions joined the legal suit.

The response in the Philippines was more muted and many commentators pointed out the weaknesses in pressing the case. Foremost among these weaknesses is the admittedly low standards of many medical schools and the recent scandal involving cheating in the nursing board exams. Government board passing rates for doctors (25%) and nurses (49%) indicate the poor standards of professional training in the country (Medical Board Examinations 2009).

The increasing professionalization of overseas workers is reflected in the preference for choosing courses that facilitate employment abroad. Hence local standards become relevant for the employability of overseas Filipinos. According to the Department of Health (Mayen 2007, 12), 85% of health professionals are working abroad. This has led to a crisis in local health care. To compound the problem, many of these medical workers come from rural areas, already poorly served by the health profession. "The health care delivery system in the Philippines has gone critical, almost desperate" (Mayen 2007). According to Duque, Secretary of Health, for every one hundred health professionals, eighty-eight have left in search of high-paying jobs outside the country. To make the situation worse, fewer medical graduates are passing the licensure examinations. This is undoubtedly due to the general deterioration of Philippine education. According to the Department of Health, the majority of government-employed doctors who left for abroad had previously converted their medical degrees into nursing. There are more Filipino nurses abroad than in the Philippines.

Most Americans who have had dealings with medical workers have come across Filipina nurses or doctors and sometimes even Filipino specialists. Filipinos constitute the largest sector of foreign health practitioners in the U.S. They remain a vital and necessary element in America's health sector. Filipinos have long wished to be part of this industry and scores of students chose to study nursing in the hope of practicing in the U.S. Because licensing to practice as a doctor in the U.S. is very restrictive, many Filipino doctors return to school to take up nursing in order to go abroad. The local health sector in the Philippines is in a state of crisis, mostly because of the departure of its health professionals.

CASE STUDIES

Case Study (1): Established Migrants

In 1977, Annie, a midwife from Zamora in Northern Luzon, decided to try her luck overseas. Accompanied by some friends, Annie went to Spain, where she worked for two years. She then moved to London for a year and from there managed to obtain a visa to enter Canada. On the suggestion of a friend, who had preceded her, Annie decided to go to Toronto. Most of Annie's decisions were made on the basis of advice offered by close friends from Zamora, her natal village. Soon after she arrived in Toronto, Annie met her future husband, Arthur, also an Ilocano who had migrated to Canada some years earlier. Annie's Zamoran friends who had accompanied her to Spain soon joined her in Toronto.

In the succeeding thirty years (1977–2007), a growing number of Annie's and her friends' relatives have settled permanently in Toronto. On arrival, many of them obtained employment through the recommendations of Annie's group, who were working in Jewish retirement homes, hospitals, small businesses and private homes. The network among the Jewish community in Toronto has proved an excellent strategy in providing employment for Zamorans. It has also provided them with basic accommodation as well as the necessary information regarding their rights as recent migrants. Many of these Jewish families were migrants themselves and readily appreciated the problems of settlement. Moreover, they also quickly realized the advantages of establishing relations of reciprocity with members of the growing Zamoran community. These ties primarily involve labor services whose essential components are personal trust, responsibility and amiability. Hence, apart from working in private homes, looking after children and taking care of the aged, Zamorans who work in small Jewish businesses often accompany their employers on holidays or look after their employers' homes when they travel abroad.

Zamorans easily relate to their Jewish employers on the basis of personal reciprocity. Jewish and Zamoran differences are commensurable. What puzzles them are the impersonal attitudes of other Canadians. Zamorans in Canada cope easily with cultural diversity but have more difficulty in accepting cultural difference. Their Jewish employers are clearly different but they can relate to them personally. Other Canadians appear to favor impersonal relationships over personal ones. This cultural difference Zamorans find harder to comprehend.

When asked why they prefer Canada, Annie and her sister point out the relative autonomy they exercise in Toronto as opposed to the close scrutiny of local Zamoran society. Annie has two children and in the village they would fall under the disciplinary gaze and control of senior kin. In Toronto, she has almost exclusive control over their behavior. In her own house, Annie makes it clear to everyone that this is her domain, which is partly the reason why her parents did not stay too long. They complained that staying in Annie's house was like living in a factory—chores had to be done at specific times.

In general I have always been struck by the varied reasons informants have for migrating. Some go in search of better employment, others go to join family members and a few confess their interest in exploring the world outside Zamora and the Philippines. While most Zamorans in the Philippines readily complain about the economic and other difficulties of local society, they rarely express a rejection of its main cultural form. This involves establishing close reciprocal ties with significant others, leading to an intense local sociality. The content and substance of this reciprocity varies considerably, and may, of course, be conducted elsewhere.

Back in Zamora, Egdon's mother, while missing her children, shows no haste in joining them. As a teacher and senior member of Zamoran society, Estefania quickly points out the disadvantages for her of life in Canada. Apart from the bad weather, her life would consist of looking after children and other domestic chores in the isolation of a Canadian home. Estefania would have to earn her status instrumentally.

Anderson (1998) has argued that modern communication not only facilitates but also often radicalizes migrants' political imaginations. They are able to keep in close touch without having to confront the practical exigencies that, in their homelands, often require compromise. Presently

Zamorans in Canada have ready access to local Philippine news, using both the Internet and community TV programs. In a sense these migrants suffer from a double imagination. They are subject to competing sets of counterfactuals, one arising from their present situation, another from past recollections.

In 1992, shortly before visiting Toronto, I was in Zamora collecting genealogical data. Members of the revolutionary New People's Army had attacked a nearby village and several Zamorans were killed. The next day I departed for Toronto intending to provide the details of this unfortunate event. To my surprise, my friends were better informed than I was. In 1995, I again visited Toronto and for the few weeks of my stay, I was taught to do the *macarena*, a Latin dance style sweeping Canada. Some months later, back in Zamora, I commented on how well the locals danced the macarena and was informed that kin who worked in Spain had taught them. They, in turn, had passed it on to their Canadian visitors a year later. These examples should remind us that globality is a two-way process, and its direction should not be easily assumed. More recently, apart from passing on dance crazes from the village to the metropolis, Zamorans in Toronto eagerly awaited videos of dubbed Mexican and Korean melodramas sent to them from Zamora (Pertierra 2002).

While communicative exchanges are now greatly facilitated, the description above has not changed significantly, even if more information is now easily accessed from the Internet and personal blogs. Generally, Zamorans in Toronto have not had fundamental transformations as a consequence of the new communication media.

Case Study (2): Separated Families

Ramon and Imelda have three children (Melanie, Paul and Michael). Ramon worked as a hospital orderly in the Philippines and after some years decided to work abroad. He has now been away fifteen years. His oldest child was 6 years old and the youngest one when he left. Ramon is lucky. He works for an oil corporation in Saudi Arabia and has yearly paid holidays. Imelda has herself worked as a domestic worker abroad several times but has had considerably less luck in finding a good employer. She has had to return without finishing her contract several times, involving great expense and thus little savings. Imelda conceived a child outside marriage, which she raised as part of the family until Ramon found out and they separated. He provides support for his three children while Imelda, although less regularly, supports her illegitimate son.

Besides visiting for a month each year, Ramon stays in touch with his children, initially through expensive telephone calls and letters, but more recently using e-mail and the mobile. Apart from the increasing frequency of communication, the content remains basically the same. They exchange news about daily activities and family affairs. Ramon is seen as a strict

and conservative father and the children generally avoid controversial topics. When asked what they miss most about Ramon, the children generally reply that they miss the intimacies of everyday interaction. However,
during his yearly visits the household is visibly tense as the children avoid
upsetting their strict father.

Melanie (21 years), the oldest daughter, recently married and now lives in
Sydney. It's been two years since she last saw her father. The three children
remained close while their parents were abroad and have learned to rely on
one another. Despite having large extended families on both sides, Melanie
and her brothers have preferred to remain relatively isolated from their kin.
Mostly they function as a nuclear household whose head is absent.

Melanie and Paul (19 years) communicate more regularly than either
does with their parents. Paul is in his final year of nursing and hopes to
work abroad as soon as possible. The full advantage of CMICT is best
used between siblings, exchanging family news as well as personal experiences. Michael (16 years), the younger brother, does not communicate
much directly with his parents or his sister, leaving it to Paul to convey
the necessary news. Macmac, their half-sibling, now lives in the province
with his mother's sister. He sends occasional text messages but visits during
school holidays.

The mobile and the Internet are significant but not essential tools in
maintaining family relationships. However, some users, such as Melanie
and Paul take full advantage of CMICT. This example indicates that while
CMICT is significant for families with members overseas, the role this
technology plays varies from really crucial to merely supplemental. Its use
and importance depends on factors such as access, familiarity and personal
interest. Some young informants claim that as long as support is forthcoming, their parent's absence is not seen as a problem. Rather than a sign
of indifference, it mainly reflects the fact that Filipinos often have parent
surrogates or do not require emotional closeness as long as they are guaranteed structural support.

Case Study (3)

Norma is a woman in her fifties who lives in a provincial town in Batangas.
Her husband has been working in Saudi for the last twenty-five years, coming home every year for holidays. She has raised her three children alone,
with the remittances and long-distance emotional support of her spouse.
She lives in a compound surrounded by her large extended family, friends
and acquaintances—but she feels lonely.

She keeps in touch with all her contacts by means of two cell phones,
a landline phone and a broadband Internet connection at home, recently
acquired. This turned out to be a life-saving technology for her, when
her eldest daughter found a job in Riyadh and settled there a few months
ago. The loss was intensely felt, but the daughter is still present at home

every day, even if only on a computer screen. She opens her Messenger window as soon as she gets to her office in Riyadh, and her mother does the same in Batangas. The two women can observe each other, interact and share their experiences during the whole working day. Even if distance is not eliminated, a new experience of closeness has come into being (Pola 2008).

This last case is typical of more recent overseas workers whose members make fuller use of CMICT. A feature of such usage is the often inconsequential nature of communication. Just as Filipinos send apparently trivial texts, families overseas often exchange banalities. It is the exchange rather than their content that is important. When asked why they bother to send these messages, Filipinos answer in Tagalog, "nagpaparamdam" (a reminder to respond).

CONCLUDING REMARKS

What role does CMICT play in the context of diaspora, expatriate workers and the globalized homeland? Our results for the importance of direct communication in local communities limit the effectiveness of CMICT. However, technologically-mediated relationships are increasingly more common among Filipinos and hence CMICT may be expected to play a more significant role even in local communities. Thus, locality may itself be cosmopolitanized in the context of a globalized homeland.

The examples of Janna and Genuine indicate how online relationships and physical sites such as the Internet café are taking over many of the traditional functions of face-to-face relationships. Filipinos are reported to be the world's greatest users of social networking sites (Casiraya 2008) and the world's highest senders of text messages (Pertierra 2006). These are mostly banal greetings meant to convey solidarity rather than information. As Ling (2008) has recently argued, they mainly serve ritual purposes under conditions of social change.

CMICT is particularly attractive for Filipinos and other Asians who wish to extend existing networks to include new members. Friendster and other similar sites are extremely popular among Internet users. This need to connect is as much a cultural as a technological imperative. But this interest is paradoxical because many Filipino users do not provide personal information on their sites or limit its access to close friends. These networks are simply extensions of personal friendships rather than self-advertisements directed at strangers. The notion of an anonymous readership is not well developed among Filipinos whose understanding of the audience is generally local. Similar comments apply among Japanese users of social networks (AP 2008). Japanese users rarely post personal information on their sites, preferring instead to include pictures of their pets. This paradox of going public while remaining private expresses the aporia of the stranger within a network of familiars.

However, foreigners are often surprised at the quick intimacy of replies encountered when they are allowed to enter Filipino sites. This is because the interlocutors immediately assume a personal relationship rather than one more suitable for polite discourse with strangers. When Filipinos meet, the first thing they do is exchange mobile numbers. This ritual is quickly followed by texts of an intimate nature. A discourse of politeness common among middle class Westerners is largely absent among Filipinos who are used to frank exchanges among friends and kin.

Most information now reaches us via the old and new media. Even fictive television programs such as *Desperate Housewives* may be seen as sources of (mis)information. The response to the perceived slur against the competence of Filipino health workers was naturally felt much more strongly by American-based Filipinos than by their local counterparts. These professional expatriates were angered by a remark against their original homeland, while those in the homeland shrugged it off as excessive zeal. As Secretary Duque pointed out, the real scandal is not the slur against Filipino doctors working in the U.S. but rather their abandonment of the local health sector.

As expected, the effects of CMICT among diasporal subjects vary depending on the situation of use, resources available and personal interests. For established migrants such as Annie, having made the choice of permanently living in Toronto, CMICT is only a supplementary tool for maintaining strong links with her natal village. The facility and regularity of air travel and older media such as the telephone, letters, videos and television provide her with adequate connections to Zamora.

Ramon and his family remain in close communicative contact even if not emotionally close. Their separation is seen as temporary even if it has lasted for most of the children's lives. CMICT has definitely facilitated communicative exchanges between family members even if it may not have transformed these relations significantly. An exception may be the case between Melanie and Paul, who maintain a close relationship despite living far apart. Their respective social lives predispose them to use this technology more fully, while their younger siblings are content to use texts (SMS).

The case of Norma and her daughter present a more telling example of the possibilities of this technology for mimicking face-to-face exchanges. Other informants also report using the Internet almost as routinely as watching television. However, this facility is not commonly available for most Filipinos. For them, texting provides the major technology of communication.

A recent news item (Ramos 2008) pointed out that many funeral homes in Manila now provide a broadband service that allows relatives abroad to access and participate in mortuary rituals through a Web site. The guilt felt by overseas relatives unable to attend the funeral services of close kin is significantly reduced through this mediated but nevertheless actual participation in the rituals. What better example of connectivity is there than this ability to connect with the dead from abroad?

REFERENCES

Anderson, Benedict. 1998. *The spectre of comparisons: Nationalism, Southeast Asia and the world.* London: Verso.

Appadurai, Arjun. 1996. *Modernity at large: Cultural dimensions of globalization.* Minneapolis: University of Minnesota Press.

Associated Press. 2008. Japan's online social scene isn't so social. *GMA News Online,* September 26. http://www.gmanews.tv/story/123152/Japans-online-social-scene-isnt-so-social (accessed May 2009).

Casiraya, Lawrence. 2008. RP has highest percentage of social network users—study. Inquirer.net. http://newsinfo.inquirer.net/breakingnews/infotech/view/20080508-135336/RP-has-highest-percentage-of-social-network-users----study (accessed May 20, 2009).

Castells, Manuel. 2001. *The rise of the network society.* London: Blackwell.

Chen, Yeong-Shang. 2004. Virtual spaces for imaginable marriages. *Asian Studies* 40(1):35–62.

Constable, Nicole. 2005. Love at first sight? Visual images and virtual encounters with bodies. Institute of Ethnology, Academia Sinica, Taipeh.

Fortunati, Leopoldina, James Katz, and Raimonda Riccini. 2003. *Mediating the human body: Technology, communication and fashion.* London: Lawrence Erlbaum.

Gupta, Akhil, and James Ferguson. 1997. *Culture, power, place: Explorations in critical anthropology.* Durham, NC: Duke University Press.

Hannerz, Ulf. 1993. When culture is everywhere. *Ethnos* 58(1–2):95–111.

Josgrilberg, Fabio. 2008. A door to the digital locus: Walking in the city with a mobile phone and Michel de Certeau. *Wi: Journal of Mobile Media* 7(Spring).

Kirby, Vivian. 1997. *Telling flesh.* New York: Routledge.

Kopomaa, Timo. 2000. *The city in your pocket.* Helsinki: Gaudeamus.

Kuvaja, Kristiina, and Anja Mursu. 2003. Sustainable development theories. *Pilipinas: A Journal of Philippine Studies* 40:55–72.

Lara-de Leon, Kristinne. 2007. An ethnographic study of an Internet café. Master's thesis, Ateneo de Manila University: Department of Sociology & Anthropology.

Ling, Richard. 2008. *New tech, new ties: How mobile communication is reshaping social cohesion.* Cambridge, MA: MIT Press.

Mayen, Jaymalin. 2007. 85% of RP health care professionals now abroad. *Philippine Daily Inquirer* (Manila), October 22.

Medical Board Examinations. 2009. Statistical Data of Medical Board Exam Passing Rates, Professional Regulation Commission of the Philippines.

Pertierra, Raul. 2002. *The work of culture.* Manila: De La Salle University Press.

———. 2006. *Transforming technologies: Altered selves.* Manila: De La Salle University Press.

Pertierra, Raul, Alicia Pingol, Minda Cabilao, and Marna Escobar. 1992. *Remittances and returnees: The cultural economy of Ilocano migration.* Quezon City: New Day Publications.

Pingol, Alicia. 2001. *Remaking masculinities.* Quezon City: UP Center for Women's Studies and Ford Foundation.

———. 2006. Brides in foreign castles. *Pilipinas: A Journal of Philippine Studies* 46:50–62.

POEA 2009 Philippine Overseas Employment Agency annual reports 2009.

Pola, L. 2008. Personal communication from the field.

Ramos, Marlon. 2008. Online wake. *Philippine Daily Inquirer* (Manila), November 2.

Ravindran, Gopolan. 2008. The rhizomatic flows of transnational Tamil cinema in Asia and Web 2.0. *Philippine Sociological Review* 55:64–78.

Tomlinson, John. 1999. *Globalization and culture.* Chicago: University of Chicago Press.

Turner, Victor. 1974. *The ritual process.* London: Penguin.

8 Make Yourself at Home in www.cibervalle.com

Meanings of Proximity and Togetherness in the Era of "Broadband Society"

Heike Mónika Greschke

INTRODUCTION

Transnational populations increasingly rely on the Internet to create and maintain links between sites of being and sites of belonging because it provides tools for establishing social relations across any distance. However, little is known about the implications of this fast-growing integration of Internet facilities into the everyday lives of geographically dispersed communities such as families or ethnic groups. This chapter introduces some emerging forms of global togetherness which are enhanced by the ongoing process of mutual integration of information and communication technologies (ICTs) with transnational everyday life. Drawing on data from multi-sited ethnographic research, this chapter discusses some results of a recently concluded study (Greschke, forthcoming). During the first phase of generating data from the virtual scene, the practice of interrelating local and virtual settings has emerged as a striking phenomenon within this field. In order to comprehend this multilayer techno-social formation properly, a multisited-ethnographic approach has been developed which relates the data gathered on the online discussion forum to data which was collected face-to-face in some of the places of residence of some of the participants, i.e. in Paraguay, Buenos Aires, California and Germany.[1] I introduce an "ethnic group" that inhabits a common virtual space in the World Wide Web (WWW) while being physically located in various socio-geographical contexts.

The case study consists of www.cibervalle.com,[2] an online discussion forum which connects Paraguayans from different parts of the world. Despite the initial anonymity of Cibervalle users to one another and their differing degrees of engagement, over the years the socio-electronic network has become a community based on solidarity and trust.

The social life of the community relies on specific communicative practices joining several mediated forms of communication with co-presence-based encounters. These practices allude to the emergence of global forms of living together. By means of sharing everyday-life practices with their distant counterparts, the communication in Cibervalle, despite the geographical dispersion of its members, has reached a high degree of proximity and intimacy. But how can we imagine people living together when they do not share the same geographical space?

In the following I explore the meaning of "living together" while being physically apart by exploring some of the communicative forms that were developed by the members of Cibervalle. In particular, I focus on the interactions between advancing technologies and the social practices of their usage on the one hand and the relation between co-present and mediated forms of communication on the other. By doing so, I hope not only to uncover the techno-social requirements for global communication, but to also exemplify some of the emerging techno-social hybrid forms that enhance global togetherness.

LOOKING AT THE VANGUARD OF BROADBAND SOCIETY: POTENTIALS OF THE INTERNET FOR TRANSNATIONAL EVERYDAY LIFE

The transnational lens on migration research, which has first been adopted from Basch, Glick Schiller and Szanton Blanc (1994), has called into question the meaning of basic concepts such as integration and belonging while highlighting social processes and institutions that have been mostly obscured by former scholarship. Instead of conceptualizing the fields of research from the framework of host or home states, transnational researchers have been studying migration processes by following the pathways of their subjects. By doing so, they have illuminated the manifold ways in which contemporary migrants act and move within so-called transnational social fields (or spaces) which span at least between places of origin and destination. As a consequence, Glick Schiller and Levitt (2004) highlight the analytical distinction between "ways of being" and "ways of belonging". "*Ways of being* refers to the actual social relations and practices that individuals engage in . . . In Contrast, *ways of belonging* refers to practices that signal or enact an identity which demonstrates a conscious connection to a particular group" (Glick Schiller and Levitt 2004, 606). The authors stress that ways of being often do not align with those of belonging, whether in terms of everyday-life practices or of political and social activities. One might have many contacts with relatives or friends in the locality of origin (being) without feeling like a member of the community at that place in ethnic or national terms (belonging). On the other hand, one might use symbols and enact practices that signal belonging to the host society (e.g. flying the host state's national flag during the World Cup) while missing the legal rights and resources to enact practices of being an equal member of this society. The analytical distinction between ways of being and ways of belonging certainly allows for a differentiated view on contemporary migration processes. However, the role of media, and especially the Internet, in this complex process of positioning and identity management in the context of migration has been largely dismissed by migration research.

In 1998, Appadurai had already spoken of "a growing number of diasporic public spheres" (22) produced by electronic media which link together individuals across national boundaries. Three years later Karim (2001) still had to claim that "the application of new communication technologies by diasporic groups has largely escaped scholarly scrutiny" (645). More recently, the new media has been addressed in some studies in the field of transnational research which seek to illuminate various diasporic public spheres. Accordingly, the potential of the Internet for political empowerment and a way of negotiating ethnic or national identity has become the focus of these studies (Graham and Khosravi 2002; Panagakos 2003; Adams Parham 2004; Bernal 2006).

But not every person involved in migration, or having transnational ties and practices, necessarily engages in politics. Concerns about belonging to a specific ethnic or national group are furthermore just one possible aspect of the everyday life of transnational populations. Besides being a technological tool for creating public discursive spheres, the Internet consists of an advancing set of means of communication that is used increasingly for creating and maintaining social relations at a distance.

Referring to the increasing use of e-mail for maintaining relationships in dispersed social networks, Georgiou (2002) stresses the significance of sharing everyday-life experiences for maintaining a sense of community. Miller and Slater (2000) assume that the Internet constitutes "an inexpensive way not only for families to be in touch, but to be in touch on an intimate, regular, day-to-day basis that conforms to commonly held expectations of what being a parent, child or family entails" (56). The authors explore how the specific use of the Internet alters senses of love and attachment, family life and other kinds of intimate social relationships. Their results demonstrate how the Internet facilitates the reintegration of the Trinidadians' basic social institution, family, which at first had been challenged by mass emigration. Wilding (2006) also examines how and whether kin maintain contact across time and space by comparing the practices of ICT usage of migrants in Australia and their relatives in six different countries. It is clear from this research that the availability of ICTs depends much upon the social and cultural contexts of family life. Furthermore, the author stresses factors like past histories and present dynamics of family relations, as well as purpose and content of communication, all of which significantly influence the use of ICTs in transnational families.

As these studies demonstrate, the Internet turns out to be a convenient tool for creating links between sites of being and sites of belonging, as it allows for the maintenance of formerly locally embedded intimate relationships which were formed on the basis of co-present interaction.

In the case of Cibervalle, however, kinship between the counterparts is rare. To the contrary, users from all over the world incorporate themselves into the socio-electronic network even though they do not have any personal relations at all within the group. Nevertheless, over the years they have made themselves comfortable within Cibervalle, thereby turning the formerly anonymous socio-electronic network into a close-knit community which is based on solidarity and trust.

www.cibervalle.com: A PARAGUAYAN PLACE OF BELONGING AND BEING

Cibervalle.com is an online discussion forum that connects Paraguayans from all over the world. In contrast to the existing body of literature on diasporic mediascapes (Karim 2001; Adams Parham 2004) and Internet usage amongst transnational populations (Uimonen 2003; Graham and Khosravi 2002), which tends to focus on common political interests as starting points for virtual representations, the Paraguayan online discussion forum started neither with a thematic focus nor with a common political goal. The beginning of this community was an electronic bulletin board which was part of the technological environment of a commercial Paraguayan Web portal. One of the principal users explains:

> The first time I entered the forum there were only five users. I have the number 6 haha. There was nobody, not even a hair moved. I began to invite friends, that's how the forum began to grow. (Manuela, Asunción, e-mail interview, *my own translation*)

Over the years, the forum has been accessed by increasing numbers of people (predominantly Paraguayans) living both at home and abroad. At the time of initial registration, the vast majority of participants of the online forum do not know each other personally. Belonging to Paraguay, however, appears to become a functional category of social organization and orientation within the global framework of cyberspace that brings together the needs of Paraguayans abroad with the desires of those wanting to leave the country and the interests of compatriots remaining in Paraguay.

The meaning of Cibervalle differs depending on the users' socio-cultural context: for those accessing Cibervalle from Paraguay, the electronic network provides a type of access to remote cultures and areas of the world which are otherwise inaccessible to them for economic reasons. This idea of virtual mobility, however, is exclusive in Paraguay because the cost of Internet access is among the highest in Latin America. Accordingly, the users located in Paraguay are part of a small affluent class defined by higher education and professional employment, both of which enable free access to the Internet. Expatriate Paraguayan users often refer to Cibervalle as a "window to Paraguay" through which they not only keep themselves up-to-date regarding current political and social issues, but also keep themselves involved in their socio-cultural context of origin on a regular, intimate and day-to-day basis. Although the degree of engagement differs from one user to the next, over the years the formerly anonymous socio-electronic network has become a central part of the everyday lives of users in Paraguay and abroad.

The community's social life relies on cross-media architecture, which joins the WWW-based public site and more privately mediated forms of communication (e.g. instant messaging and mobile phone use) with co-presence-based encounters. Cibervalle, in other words, has become a shared social space which spans physical, grounded localities in some Paraguayan cities, Buenos Aires, New York and other places all over the world, and it turns out to be both a site of *belonging to* Paraguay and of *being with* Paraguayans at a distance. The construction of Cibervalle as a place for meeting friends, joining events and sharing everyday life is a practical achievement relying on evolving communicative practices and technological features. As Bergmann and Meier (2003) argue, social sciences should be particularly concerned with the manifold processes of the new media's integration into everyday life, in order "to describe the forms of interaction and communication that they create and to show how they influence, shape, create or hinder processes of society-building" (430, *my own translation*).

Consequently, I reconstructed the process of mutually advancing technological tools and their usage for the time from 2002 to 2005. For this purpose I used an Internet Wayback Machine that enabled me to find older versions of the electronic bulletin board and to examine the Web site renovations made over time. Via the ethnographic data, I determined that the technological development of the electronic bulletin board was strongly shaped by the user's needs and communicative practices. Furthermore, through my own participant observation I was able to grasp the meanings of those technological renovations. In the following section the results of my analysis, which focus on those interrelations between advancing technologies and the users' practices of everyday appropriation, which I assume to be crucial for the evolution of global togetherness, will be discussed.

FROM AN ANONYMOUS ELECTRONIC NETWORK TO COMPUTER-MEDIATED COMMUNITY LIFE: CIBERVALLE AND THE EVOLUTION OF GLOBAL TOGETHERNESS

Once upon a time there was . . . just one more electronic bulletin board, entered by many nicks in order to read, to argue, to maunder, to calm down, to forward poems, to learn tags, to joke, to write novels and tales, to teach sex education, in short. . . I wanted to be part of that group, I read them for about three months and started to imagine how each one of them would be in reality. At that time there was neither a photo album nor member profiles, and hardly anyone had proficient usage of html, and we knew even less how to upload photos. Writing to someone in particular seemed like a shot into the darkness, because you never knew whether and how he or she would answer. To imagine the smile, voice or face of your counterpart was utopia. Knowing his or her strengths and weaknesses you could not even dream of. (Esther, Paraguay, Cibervalle forum, *my own translation*)

This comment builds part of a discussion thread that reconstructs the history of Cibervalle. Esther is one of the long-standing members, who have actively been shaping the evolution of computer-mediated communication over the last few years. During her membership she has been involved in several technological renovations of the electronic bulletin board. Stressing the condition of anonymity as much as the scarce means of expression in the beginnings of Cibervalle, she points toward the connection between informational richness of communication and the quality of social relationships.

The Online Forum

The development of computer-mediated communication was still in the fledgling stages and broadband technology was not yet available when the first Paraguayan Internet travelers began to settle down in Cibervalle. The technological requisites were still insufficient for establishing personal relationships. Although Esther stresses the limits of personal relationships due to the scarce communication modalities, her comment also hints at the means through which the Cibervalle community has been exploiting the potentials of the Internet. She specifies some initially lacking features (photo album, member profiles) and skills (computer literacy, like HTML) which have been mutually acquired by both users and technology. The renovation of the system infrastructure from 2002 to 2005 aimed as much at the enrichment of the informational thickness of communication as of the increase in the variety of topics. Furthermore, they reflect changing modes of temporality.

Enriching Informational Thickness of Communication

At the beginning, the communication on the Cibervalle forum was limited to the exchange of unformatted text. Some of the more computer-savvy members then began to use HTML tags to enrich their communication by making use of different colours, font sizes and visual elements. The HTML tags enabled the members to express emotions, for instance by using loud colours and big font sizes. During the following technological renovation, the operating company adopted these practices by augmenting the forum with a user-friendly input mask which facilitated formatting text and adding elements and links. The ability to add visual elements or upload photos corresponds to a desire of overcoming the constraints of anonymity. Besides photos of the members themselves, users also shared visual impressions of their physical environments.

Enlarging the Variety of Topics

Esther's comment points to another important feature of the Cibervalle forum, namely the wide range of topics and activities taking place there. Unlike other online discussion forums, this Paraguayan virtual community does not limit the thematic focus of postings. In fact, most of the activities in this forum reflect shared everyday-life concerns: talking about the film one saw last night, discussing where to get the most authentic ingredients for preparing *sopa paraguaya*, telling jokes and laughing together, inventing forum games and playing together, teasing each other, quarreling, debating forthcoming elections, exchanging opinions about homosexuality, abortion, religious beliefs, sharing happy or difficult moments in life, discussing concerns about the future perspectives of Paraguay. In short, the forum reflects a wide range of topics and socializing activities that usually take place in families at lunch time, or among friends who meet each other on the *plaza central*, i.e. the town square. Many of the users consider Cibervalle a central part of their life-worlds, where a significant part of their socializing activities take place.

The evolution of the Cibervalle forum also reflects the basic orientation of members towards everyday life and socializing rather than on particular topics. The thematic order of the forum was originally organized into sections such as football, politics and general topics in 2002, and it gradually expanded over time to incorporate a variety of topics in up to nine different sections. However, the members do not seem to be concerned with the proposed thematic orderliness. Rather, they appear to organize communications following their own practical logic that is connected to modes of temporality, as will be explained in the next section.

Changing Modes of Temporality: Synchronicity and Asynchronicity

Unlike other Internet-based communicative genres like chat or e-mail, the discussion forum provides temporal freedom and the choice between synchronous and asynchronous communication. Users who are online at the same time are able to communicate with one another in real time. However, those who do not spend much time on the Internet or are not able to be online at the same time as their counterparts are still able to reconstruct ongoing discussions and make contributions. This possibility arises because the discussion threads remain displayed and the postings can be read in their chronological order. In brief, the condition of asynchronicity allows for global communication between actors who live in different time zones (Stegbauer 2000, 25). Synchronicity, on the other hand, leads to time and space compression, and enables real time communication regardless of the distance between counterparts. In the case of the Paraguayan online forum, it seems as if these features have provided the technological conditions for the development of a global community. The forum bridges temporal and geographical distances. It is able to unify people despite different degrees of involvement and may hence create a sense of togetherness, even under the condition of geographical dispersion. Over time, however, the practices of the participants have been leading to some changes on the electronic bulletin board which reflect the fact that communication among members is structured temporally rather than thematically. The members who have permanent Internet access at home or in their work place normally enter Cibervalle in the morning and review the current activities. They then become involved in daily tasks in their physical environments while leaving the Web site open. Periodically, they approach the computer and push the refresh button in order to update the news from Cibervalle.

Due to the everyday routines of the members, the list of currently debated threads, the so called *Recientes,* has become the most important part of the Web site. The *Recientes* displays a selection of the currently debated threads in chronological order from newest to oldest, where the date is taken as the date of the last post. When a member posts in a thread it will jump to the top because it is the latest updated thread. Similarly, other threads will jump in front of it when they receive posts. In 2002 the *Recientes* contained no more than the thread's title and its author's nickname.

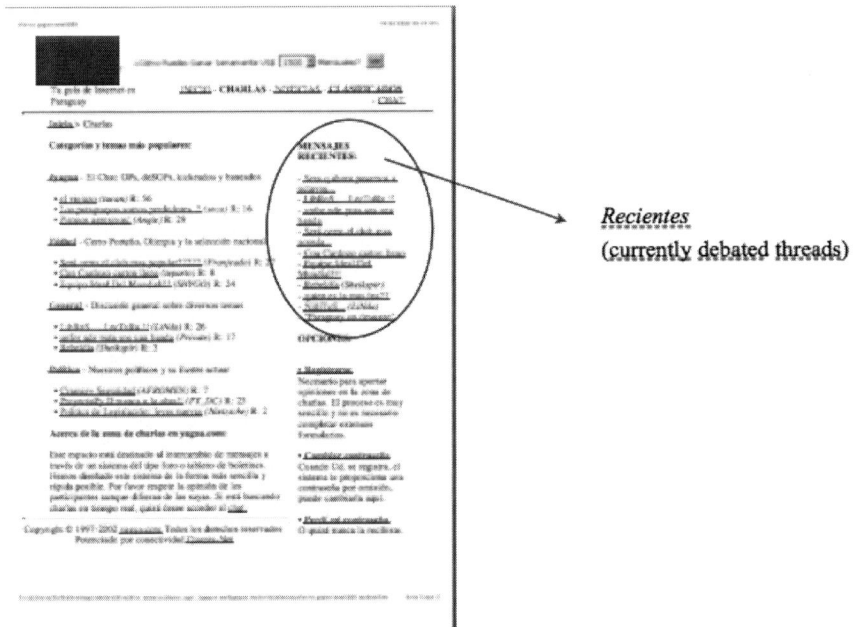

Figure 8.1   Cibervalle forum 2002.

Over the years, however, the *Recientes* has been delivering more information about the activities in Cibervalle. In 2005 the *Recientes* began reporting the national context of message authors, the number of times a thread was viewed and the number of replies to a thread.

"What is happening at the moment in Cibervalle?" seems to be the guiding question. The *Recientes* appears to be the place where the forum movements become visible and observable. With the *Recientes* one can experience Cibervalle's social life on its own, independently from one's own active participation.

Instant Messenger

The members of Cibervalle do not only communicate via the online forum. For more private chat, as well as for specific requests that need to be discussed instantly, they also use instant messaging. Instant messaging (IM) defines a program that enables instant communication between two or more people over a network such as the Internet. Most programs can be downloaded from the Internet at no cost. The IM program used by most of the Cibervalle members enables one to add and remove the e-mail addresses of contacts one likes to network with. After logging into an IM program, a window appears (the so-called buddy list) which shows the names of contacts who are currently online. When clicking on one of the names, a second window appears (the private conversation window) which is used

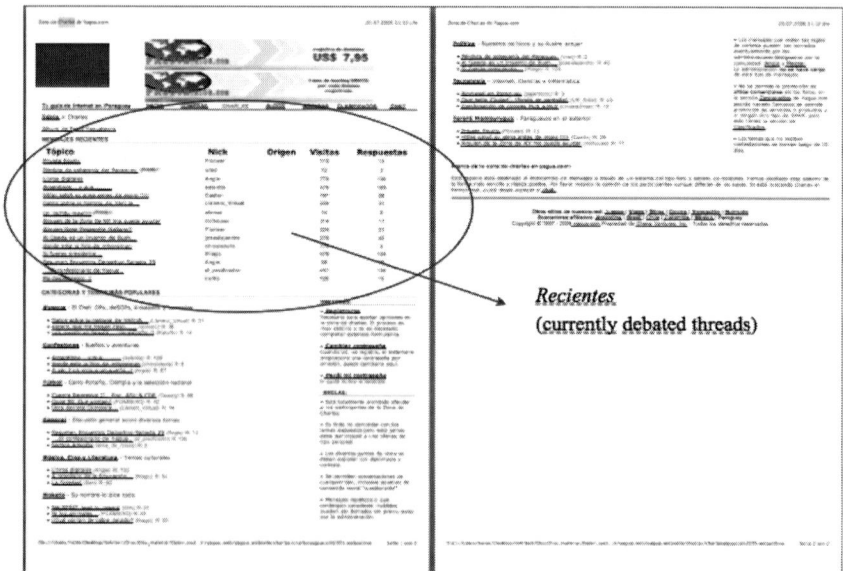

Figure 8.2   Cibervalle forum 2005.

to send instant messages to an online user. On the screen of the addressed user another window appears displaying the message that was sent. Besides the visual information, the program emits a sound every time a message is received or one of the contacts logs in. IM is also used for discussing current topics "in private" and events which both counterparts can follow simultaneously on the discussion forum. The communication displayed in the public forum thus tends to also be influenced by private conversations. By combining different technologies, common interests are produced, which in turn constitute a shared social space.

Entering Cibervalle is like opening up an additional techno-social space with globally shared common interests that overlay the users' local environments and which have to be coordinated with the socially relevant details on the spot. The practice of keeping the page open and refreshing the screen from time to time may enhance the imagination of approaching and distancing oneself (but not leaving) a social space created and maintained by the communicative activities of its users. This virtual space and the social activities within it, however, go on independently of the actual participation of the individual user. Chatting with other members via IM about what is going on in Cibervalle as well as joint travels through the online forum may even strengthen the perception of sharing the same space. Both instant messenger and the online forum enhance the sense of being together even if the counterparts are actually separated by thousands of miles.

At the first glance, it seems as if computer-mediated communication has been superseding co-present encounters. The Internet technologies in use enable the users to be in touch on an intimate, regular and day-to-day basis not only with distant friends and family members but also with the social life going on in their place of origin. The communication within the shared virtual space seems to be a functional equivalent of being physically together in the same place. In fact, this spatial formation is structured through timely coordination rather than in geographical terms. This insight is not as trivial as it appears at first glance. The alignment of synchronous communication within Cibervalle does indeed lead to the differentiation of the network into subgroups uniting those members who are frequently online at the same time. In other words, for creating and maintaining intimate relationships it seems to be more important to share the same time zone than the same geographical place. Nevertheless, geography still matters in another respect, as the following statement indicates:

> Once I was bored sitting in front of the screen of my pc and I thought: Internet, Internet, my boyfriend says there is everything for anyone, "ndeee" imagine that! And me, what I am going to do? Very well then google: 'Paraguay' the first thing to appear was cibervalle. com. I entered, registered myself and began to follow the discussion threads, sometimes I posted some comments and one day I decided to post a thread of my own . . . I came to know some of the Argentinean

members, I went to an encuentro and that's how I had my first meeting with Paraguayans in Argentina. Until then, I had never met any
Paraguayan here during the whole four years I had been in Argentina.
It was a very good experience and I turned to re-appreciate my culture
as well as my country. (Iwashita, Buenos Aires, Cibervalle forum, *my
own translation*).

Iwashita's comment reveals the meanings of geography in Cibervalle.
As a Paraguayan migrant she used her national affiliation for orientation
through the World Wide Web ("very well then google: 'Paraguay'"). At
the time, Cibervalle was one of the most important Paraguayan Web sites
and it ranked first in results from the Google search engine. Indeed, many
of the expatriate participants came across the Cibervalle forum when they
were searching the Internet for information about Paraguay. Furthermore,
Iwashita names typical stages and elements of Cibervalle membership. Like
most of the members, she observed the online activities for a while in order
to become acquainted with the topics and practices before she introduced
herself as a new member. As part of the welcome reception she was then
asked by the other members to reveal her actual place of residence. Then
the Argentinian members invited her to their next "encuentro", where she
finally met other Paraguayan migrants living in Buenos Aires. Again, place
matters. The place of origin, on the one hand, serves as a shared point of
reference within the virtual space, the final frontier. On the other hand,
based on the place of residence, the global community divides into local
subgroups which are different from the ones based on time zones.

Local Meetings

The virtual activities in Cibervalle are, admittedly, just one side of the coin.
This is because the members regularly meet each other face-to-face. Locat-
ing each other in order to meet personally in fact is one of the fundamental
practices I was able to observe in this field. The forum provides the chance
for networking with compatriots who may currently live at the same geographical location or nearby, not only for emigrated Paraguayans. For both
those at home and abroad, the virtually formed relationships seem to stimulate the need for face-to-face encounters. By organizing *encuentros*, which
are regular meetings in their respective localities of residence, the virtual
relations are translated to local contexts whenever possible.

By exploring the particulars of mediated vs. co-present communication,
Urry (2003) demonstrates the significance of co-present encounters for creating and maintaining social networks. As face-to-face—or body-to-body
(Fortunati 2005)—encounters afford access to the eyes of the other, they
are considered the most direct and reciprocal form of interaction. "Eye
contact enables and stabilizes intimacy and trust, as well as the perception of insincerity and fear" (Urry 2003, 164). The informational thickness

of physical co-presence is reached because "conversations are made up of not only words, but indexical expressions, facial gestures, body language, status, voice intonation, pregnant silences, past histories, anticipated conversations and actions, turn-taking practices and so on" (Urry 2003, 164f). Besides the embodied particulars, Urry highlights place and time aspects of co-presence. Sharing the same location in a specific moment and experiencing a place with one's own senses, "physically walking or seeing or touching or hearing or smelling a place" (Urry 2002, 261), is considered to be a crucial base of co-presence and explains the increasing need for physical travel enhanced by the spread of global means of communication.

Nevertheless, in Cibervalle the face-to-face relationships, initially enabled by the electronic network, by no means substitute for the virtual ties. Rather the local subgroups which have been evolving along the way in different parts of the world share the events they organize at the local levels afterward with the global community. That is to say, soon after a local meeting occurs, the group reports about it on the electronic bulletin board using images of the participants and their activities. The distant members of the global community respond, expressing gratitude for sharing the event with them, and they usually add some charming or funny comments to the images. In so doing, the events that are first enacted by local subgroups are reconstructed on a global level as a collective experience of the whole community.

In this regard, I want to highlight one of the main findings of my research, namely the fact that the sense of the local events seems to be strongly shaped by the delayed reconnection with the virtual level. Particularly, the meetings in Paraguay may be taken as performances of everyday life in Paraguay that are addressed to the people abroad. This might explain why the users abroad perceive Cibervalle as "a window to Paraguay", through which they keep themselves involved in their socio-cultural context of origin on a day-to-day basis. With Cibervalle, Paraguayans in Spain, Japan, the U.S., France, etc. find an easy and inexpensive way to take daily trips to Paraguay, irrespective of the actual geographical distance to their place of origin.

Again, although "virtual travel" (Urry 2002) may "ease nostalgia" and "make one feel closer to home", as some participants explain, it seems also to enhance the need for occasional meetings. Indeed, not only the members located in geographical proximity make it a habit of meeting each other personally. Whenever one of the distant members spends his or her holidays in Paraguay, the local groups organize welcome meetings for the visitor. These meetings are also collectively reconstructed and shared with the whole community. In brief, "meetingness" is assumed to be a fundamental feature of the Cibervalle community, because "moments of physical co-presence are crucial to patterns of social life that occur 'at-a-distance' . . . 'Meetingness', and thus different forms and modes of travel, are central to much social life, a life involving strange combinations of increasing distance and intermittent co-presence" (Urry 2003, 156).

Social life within Cibervalle is based on a complex relationship between different forms of communication, enhanced by specific combinations of physical, imaginative and virtual travel. The practice of interrelating local events (the meetings) with virtual events on a global level (the meeting's collective narrations) transports the members of the group imaginatively to the localities of residence of their counterparts. The possibility for daily travel to the places of one's origin is assumed to be particularly important for Paraguayans abroad, as it is hard to obtain any images or information about this country through the mass media. The specific simultaneous use of IM and an online forum enables joint travels through a shared virtual space in real time and creates experiences of proximity. This kind of computer-mediated co-presence seems to dissolve the boundary between physical presence and total absence.

CONCLUSION

In this chapter I have explored some implications of Internet usage in transnational migrants' everyday life. I have argued that although transnational research has significantly contributed to a more precise understanding of contemporary migration, the role of digital media has been largely neglected. The existing body of literature either focuses on the Internet as a tool for maintaining already existing relationships or assumes the existence of a community with common interests regarding the "homeland" as a starting point of virtual performances. In the case of Cibervalle, however, neither common goals or topics nor pre-existing personal relationships brought the counterparts together, yet it has over the years become a community of trust and solidarity for many of its members. In this process of community building, the interactions between advancing technologies and the progressive proficiency of users has been crucial.

In this respect, I have demonstrated the benefits of focusing on the interactions between advancing technologies and the social practices of usage in order to illuminate forms of interaction and sociality that have been emerging in the everyday lives of contemporary migrants. The members of Cibervalle have not only turned the formerly anonymous socio-electronic network into a community, thereby intimately connecting the electronic bulletin board with their physically embedded life-worlds, they have also become part of a mutual learning process between technology development and practice.

The specific use of Internet technologies as a quotidian means of communication in transnational migration contexts engenders "global togetherness". This does not only apply to transnational populations who already share the experience of living together and now, while geographically apart, have to maintain their familiarity—even those entirely unknown to each other and distant in geographical space are enabled by the Internet to be close in virtual space where they can learn to create the experience of living together.

In the context of migration the availability of distance-shrinking communication technologies becomes crucial in terms of maintaining *and* creating links between sites of being and sites of belonging. In turn, the complex interactions between the various means of communication and the specific communicative practices developed in transnational migration contexts seem to bring about new global types of social relations and sociality. Transnational populations, in other words, turn out to be a kind of vanguard of broadband society, as they point toward trendsetting hybrid forms of techno-sociality.

The Internet in and of itself does not have this intrinsic potential. It is the creative practices of the users who bring about the development of the Internet as a means of facilitating global togetherness.

NOTES

1. For a more detailed discussion on the methodological implications of doing ethnographic research in plurilocal and computer-mediated fields, see Greschke 2010.
2. The online forum's name and the nicknames of the users were changed as a means of preserving the participants' identities.

REFERENCES

Adams Parham, Angel. 2004. Diaspora, community and communication: Internet use in transnational Haiti. *Global Networks* 4(2):199–217.

Appadurai, Arjun. 1998. *Modernity at large: Cultural dimensions of globalization*. Minneapolis: University of Minnesota Press.

Basch, Linda, Nina Glick Schiller, and Christina Szanton Blanc. 1994. *Nations unbound: Transnational projects, postcolonial predicaments, and deterritorialized nation-states*. London and New York: Routledge.

Bergmann, Jörg, and Christoph Meier. 2003. Elektronische prozessdaten und ihre analyse. In: *Qualitative forschung: Ein handbuch*, ed. Uwe Flick, Ernst von Kardoff, and Ines Steinke, 429–437. Reinbeck: Rowohlt.

Bernal, Victoria. 2006. Diaspora, cyberspace and political imagination: The Eritrean diaspora online. *Global Networks* 6(2):161–179.

Fortunati, Leopoldina. 2005. Is body-to-body communication still the prototype? *The Information Society* 21(1):53–62.

Georgiou, Myria. 2002. Diasporic communities on-line: A bottom up experience of transnationalism. London School of Economics and Political Science. http://www.lse.ac.uk/collections/EMTEL/Minorities/papers/hommesmigrations.doc (accessed November 16, 2005).

Glick Schiller, Nina, and Peggy Levitt. 2004. Conceptualizing simultaneity: A transnational social field perspective on society. *International Migration Review* 38(145):595–629. http://www.peggylevitt.org/pdfs/Levitt-Glick.conceptsimult.pdf (accessed May, 30, 2010).

Graham, Mark, and Shahram Khosravi. 2002. Reordering public and private in Iranian cyberspace: Identity, politics, and mobilization. *Identities: Global Studies in Culture and Power* 9: 219–246.

Greschke, Heike Mónika. Forthcoming. *Is there a home in cyberspace? The Internet in migrants' everyday life and the emergence of global communities*. New York: Routledge.

———. 2010. Mediated cultures of mobility: The art of positioning ethnography in global landscapes. Working Paper 78, Centre on Migration, Citizenship and Development (COMCAD), Bielefeld. http://www.uni-bielefeld.de/tdrc/ag_comcad/downloads/workingpaper_78_Greschke.pdf (accessed July 13, 2010).

Karim, Karim H. 2001. From ethnic media to global media: Transnational communication networks among diasporic communities. *Nord-Süd aktuell* 4:645–654.

Miller, Daniel, and Don Slater. 2000. *The Internet: An ethnographic approach*. Oxford: Berg.

Panagakos, Anastasia N. 2003. Downloading new identities: Ethnicity, technology, and media in the global Greek village. *Identities: Global Studies in Culture and Power* 10:201–219.

Stegbauer, Christian. 2000. Begrenzungen und strukturen Internetbasierter kommunikationsgruppen. In *Soziales im Netz: Sprache, beziehungen und kommunikationskulturen im Internet*, ed. Caja Thimm, 18–38. Opladen: Westdeutscher Verlag.

Uimonen, Paula. 2003. Mediated Management of Meaning. Online-Nationbuilding in Malaysia. *Global Networks* 3(3): 299–314.

Urry, John. 2002. Mobility and proximity. *Sociology* 36(2):255–274.

Urry, John. 2003. Social networks, travel and talk. *British Journal of Sociology* 54(2):155–175.

Wilding, Raelene. 2006. "Virtual" intimacies? Families communicating across transnational contexts. *Global Networks* 6(2):125–142.

9 The Bulgarian-Language Media Diaspora

Polina Stoyanova and Lilia Raycheva

INTRODUCTION

This chapter discusses some major political and social implications of the Bulgarian Diaspora viewed through the prism of the Bulgarian-language media. It evaluates the role of the information and communication technologies in the process of social inclusion within the local media consumer community, as well as in the dynamics of national integration. The research for this chapter arises out of discussions at the World Meetings of Bulgarian-Language Media initiated by the Bulgarian Telegraph Agency (BTA). The chapter refers to the first six of these meetings that have been held respectively in Sofia (Bulgaria), Chicago (U.S.), the Vatican, Madrid (Spain), Varna and Bourgas (Bulgaria) and Vienna (Austria); their aim is largely to promote Bulgaria around the world. The main problems discussed at these meetings correspond to some of the most pressing issues for the Bulgarian public at home and abroad. Numerous professional discussions and cultural programs, including concerts and exhibitions, have been carried out in parallel during these meetings. These events have promoted the significance of the introduction of the Cyrillic alphabet to the European community; presented the millennia-old Bulgarian history and culture; highlighted the "You Are Not Alone" international media campaign in support of liberating imprisoned Bulgarian medics in Libya and more. Additionally, a major goal of these meetings was to present the work of the Bulgarian-language journalists (in press, radio, TV and online media) on the problems of the Diaspora and to establish collaboration between them and their colleagues from Bulgaria.

The main focus of this chapter is thus situated on the key characteristics of the Bulgarian-language media abroad. It explores data collected in a special survey designed to gather material on the Bulgarian-language media outlets and the Web-based societies linked to these media, as well as to include the monitoring of the coverage of the media forums. This survey was prompted by the need to investigate the potential of the Bulgarian-language media within the Diaspora. Methodologically, the analysis is set in a framework that explores the interrelations between the national identity and the global migration process. These interrelations are analyzed within the context of the interaction of the Bulgarian Diaspora with the forums of the six aforementioned Bulgarian-language media world meetings. This study represents new research into these problems and the analysis also refers to the national policies addressing the Diaspora.

NATIONAL IDENTITY VS. GLOBAL MIGRATION AND MEDIA MIX

In the age of widely developed information and communication technologies, working abroad for a living can no longer have the same meaning as before. The variety of texts on the global migration process offers multiple approaches to exploring this phenomenon, as national identity is currently in the process of being gradually replaced by the global cultural mix. The frontiers of the nation-state are blurred out in a "flat world", according to the concept of Friedman (2007) in which the human being only belongs to some type of national community conditionally. According to Todorov (1998) (a French sociologist of Bulgarian extraction and a leading representative of the Bulgarian Diaspora) cultural identities are not strictly national. National cultural affiliation is simply the strongest, because it combines the imprints—both in body and spirit—of the family and community and of the language and religion. A process of deculturalization, the fading out of the original culture, now occurs in our time. However, it could be compensated for by acculturization, the gradual absorption of a new culture, of which everybody is capable (Todorov 1998).

With the advent of globalization, an ever greater number of discrete societies find themselves in foreign-language spaces. Human understanding of distance, time and reality has radically changed as a result of mediamorphosis (Fidler 1997). However, communication with their nation-states has never been easier for all diasporas across the world. Now every person who has left the territory of his or her home nation-country fully comprehends that the world is a global place and that remoteness is a conditional concept. That is why the boundaries between the concepts of Nation, Citizenship and Diaspora become obscured in the contemporary world. Indeed, every time we are tempted to mention McLuhan's Global Village, we should also recall Meyrowitz's contention that the media create communities without a sense of place (1985).

In his theory of "breaking with all kinds of the past", Appadurai (1996) articulates a view of cultural activity known as the social imaginary. For him, the social imaginary is composed of five dimensions of the global cultural flow: ethnoscapes, mediascapes, technoscapes, financescapes and ideoscapes. Appadurai regards the media and migration as the two major and mutually interrelated aspects of the globalizing processes in the transnational spaces. The images, scenarios, models and stories which reach us via the mass media create the difference between migration of today and migration of the past. People who want to migrate, people who want to return and people who want to stay seldom form their plans without the help of the media (radio, TV, newspapers and the Internet). Deterritorialization creates new markets for the movie companies, media conglomerates, impresarios and travel agencies that meet the needs of deterritorialized populations for contacts with the nation-country. But the nation-country is fabricated to some extent and exists only in the imagination of the deterritorialized groups (Appadurai 1996).

According to Lash (2004), the national, political and cultural relations are declining and being replaced by global flows. These are financial, technological, informational and communication flows and flows of images, ideas, migrants, tourists and travelers. Politically, supranational and subnational institutions threaten the hegemony of the nation-state institutions. Almost all flows constitute symbolic and cultural commodities: from images, money and ideas to communications and "traveling cultures" that travel today with the migrants (Lash 2004, 54).

In the brave new post-modern world the individual is not bound by time and space anymore. Subordinated to the dominance of popular culture and to the broad access to information and telecommunication technologies, the post-modern generation does not consciously search for its identity. Furthermore, the new "hyperspace" created by global communications makes impossible the positioning of the individual within the framework of traditional frontiers. The major missions of hypermodern times now are how to preserve the feeling of individuality in the crowd and how to develop opportunities for a free choice.

One should not lay too strong an emphasis, however, on idealization and mythologizing of the qualities, characteristics and opportunities offered by the virtual and online culture. Although it is interactive by nature, and can overcome many frontiers (ethnic, national and continental) through universalization and interaction, along with this it has an alienating and restricting effect on the people by destroying, at its most unpleasant, the "live" everyday bridges of communication between friends, kin, neighbors, colleagues, etc. According to Barbier-Bouvet (2002) discussions are always happening on the Net—you can always participate in a forum. Even if used only for reading, without exchange, the Internet is mainly a "diaspora" medium: for instance, the "diaspora" of orchid-lovers. Theoretically, the Internet fulfills the dream of omnipresence; it creates the feeling of being at the same time wherever one is and elsewhere. Practically, the Internet brings close far-away persons of the same tastes, but alienates close-by persons of different tastes (Barbier-Bouve 2002). The world Net has hastened to cater to the chance for everyone to be noticed, heard or seen, to fulfill the human need of knowing that something depends on you, of becoming a chronicler of events, of offering the others information about yourself and your world as well as about what you have seen and felt in the others, about the way the others live their life, and generally about everything that has impressed you.

If we attempt, just for an instant, to step into the shoes of the émigrés abroad, we would find out that the bond with the nation-country is a conditional and variable concept. They may feel bound to the space of their home and their kin, but generally the space of the entire country, of all their fellow citizens, of the internal political, economic, social, etc. relationships would always escape them. In this situation, we assert, the mass media come to the fore.

Following Toffler (1991, 209–210) there is a new "mosaic culture" that has been created, filtered, distributed and stored with the help of the movies, press, radio and television. This offers the possibility of longevity to any work, even monumentalizing it. Within the development of culture through this prism of media evolution, television falls into post-modern times, while the new medium—the Internet—is the mark of hypermodern times. All relationships in it are hyper: reality is hyper-reality and space is hyperspace. Mass media are the "new factors of the spiritual world" which build the new cultural spaces, where the communicated message appears to reign without bounds. Accordingly, they turn the contemporary environment into a mosaic of chaotically arranged communications exercising an enormous impact on the people (Toffler 1991).

Thus, although the "flat world" platform of Friedman (2007) may contain a potential for homogenization of cultures, it offers an increasingly strong potential for the encouragement of diversity to a degree unknown before. This is mainly due to the uploading and broadcasting of data which make "globalization of the local" possible. The "flat world" platform makes it possible for us to upload our own culture to the world and the émigrés abroad can now address their nostalgia better than ever before. Indeed, the Internet encourages the diaspora societies across the world to make use of the contemporary global media networks in order to stay closer to their local customs, news, traditions and friends, irrespective of where they live. In this context globalization of the local means "globalization backwards". Instead of having a specified region covered by the global media, the local media go global and the diaspora market is broadening with the help of newspapers, TV and radio programs all entirely in local languages and broadcast via satellite on an international scale (Friedman 2007).

POLICIES ADDRESSING THE BULGARIAN DIASPORA

The Bulgarian Diaspora today is a powerful but overlooked source of patriotism and economic potential for Bulgaria. Notwithstanding those citizens who choose to leave because they like the lifestyle of their adoptive country more, there are some individuals who have chosen to tear out their roots and leave their native place for a far-away country could take advantage of their native culture—irrespective of the different environment they live in-thousands of miles away from their native home. Thanks to the possibility of reading the native papers online, of watching the satellite TV programs, of listening to the radio on the Web and of watching the daily news, the forces of particularization now seem as powerful as the forces of homogenization.

The number of Bulgarians living outside their nation-country now exceeds 3.5 million. This accounts for nearly half (7,351,234 in 2011) of those who live in the territory of Bulgaria (National Statistical Institute 2011). There are no official surveys on the number of Bulgarians living abroad, owing to the difficulties for any such census. In an interview for the Bulgarian-language site www.eurochicago.com, Bozhidar Dimitrov, a former minister of Bulgarians living abroad (and long-term researcher of the Bulgarian Diaspora), divides them into two groups. The first group constitutes Bulgarian citizens who have left the country over the last one hundred years and have formed sizable communities in the E.U., U.S., Canada and South Africa. In his opinion, the second largest group is the Bulgarians who remained abroad owing to territorial losses by the country in the 20th-century wars, or owing to the mass migrations in the 17th–19th centuries. These are the Bulgarian communities mainly in Turkey, Greece, Macedonia, Albania, Kosovo, Serbia, Romania, Moldova, the Ukraine, Kazakhstan (Dimitrov 2010). According to the data supplied by the National Statistical Institute (NSI), in 2009 migration from Bulgaria had surged strongly as compared to the two previous years, increasing over six times.[1] In our research we shall not dwell on the pre-conditions that have led to this process. Undoubtedly, the 20th-century wars as well as the developments following the fall of the Berlin wall in 1989 have been considerable factors. In the long run, mass emigration has had a number of negative consequences for Bulgaria, which could be summed up in three main groups:

- Deterioration of the demographic situation
- Reduction of the workforce
- Brain drain

For one reason or another, these Bulgarians are now in a foreign territory, and perhaps surrounded by unfamiliar customs in new environments. There can hardly be a common denominator for the way in which the émigrés can become accustomed to the established habits in their new countries, but for some there does remain a nostalgia for their native land, native language and culture that affects the existence of these Bulgarians abroad. It is suggested here that for some Bulgarians, when one belongs to a nation, it is for life. A person cannot choose either the family they are born into or his or her nation-country, which may well have emotional ties for him/her throughout their life. In this new environment the émigré may well be faced with the challenge of preserving his or her identity as a Bulgarian national. In this sense, belonging could be regarded as a primary liaison in time and space; identity could be viewed as a secondary liaison, mediated and negotiated. The retention of national identity by the émigrés is thus very much dependent on both their own will and on influences such as exerted by the power of the media.

We now return to the role of the world meetings of the Bulgarian-language media in order to understand how they should be positioned against the national policies that relate to Bulgarians abroad. The "Constitution of the Republic of Bulgaria" lacks any fundamental provisions for fulfillment of a target-oriented national strategy and policy addressed at our compatriots abroad. In Chapter 2 of this document, which refers to the basic rights and obligations of the citizens, there are only two articles regulating the commitment of the State to the citizens residing abroad:

- Article 25 (5): Any Bulgarian citizen abroad shall be accorded the protection of the Republic of Bulgaria.
- Article 26 (1): Irrespective of where they are, all citizens of the Republic of Bulgaria shall be vested with all rights and obligations proceeding from this Constitution. (Constitution 1991)

In order to address this deficiency, a "Law for the Bulgarians living outside of the Republic of Bulgaria" has been adopted only in 2000. Under Article 2 of the Law, a Bulgarian living outside of the Republic of Bulgaria is a person who:

- Has at least one ascendant of a Bulgarian origin
- Has a Bulgarian national consciousness
- Resides permanently or constantly in the territory of another state

The Law provides for the foundation of a "National Council for the Bulgarians living outside of the Republic of Bulgaria". However, the participation of the Bulgarians residing abroad in it is subject to unclear and incongruous criteria. That is why, more than ten years after the promulgation of the Law, no such Council has yet been created. This automatically renders meaningless some of the provisions in the Law that correspond to the functions and tasks of this body (Law 2000). So far the State Agency for the Bulgarians Abroad, founded on October 1, 1992, has acted as a coordinator for the implementation of the government policy concerning the Bulgarians and the Bulgarian communities across the world. The main objectives of the Agency are firstly the preservation of the spiritual heritage of the nation—its language, culture, traditions and history—among Bulgarians across the world, and secondly the creation of Bulgarian lobbies abroad so as to assist enhancing of the prestige both of Bulgaria and of the Bulgarian communities in the respective states (Agency 2008).

It was only in May 2008, at the opening of a national round table dedicated to the new Government policy concerning the Bulgarians abroad, that the then-acting Prime Minister Sergey Stanishev outlined the need for a new—and coordinated on all levels—policy of the Bulgarian State addressing the Bulgarians residing abroad. Two strategic goals have been discussed. The first one targets the qualified young Bulgarian émigrés who have left the state mainly for economic reasons. The idea is to put to a good use their knowledge and experience for the benefit of the country. The second strategic goal is a new and updated policy for accepting nationals of third countries to assist the Bulgarian economy, as well as effective regulation and control of the migration processes.

The policies of the current right-of-center Bulgarian Cabinet (elected in 2009) addressing the Diaspora include the review and modification of the normative acts, as well as development of a special program for work with the Bulgarians across the world. A special minister for the Bulgarians living abroad has been appointed and is in charge of these policies.

WORLD MEETINGS OF THE BULGARIAN-LANGUAGE MEDIA

Traditionally, the Bulgarian-language media world meetings are organized around May 24th, the Day of Cyril and Methodius (the creators of the Cyrillic alphabet, the official alphabet used in Bulgaria), in places with a substantial number of Bulgarian residents and Bulgarian-language media. Over 250 representatives of the Bulgarian media across the world have taken part in the six editions of this initiative. The meetings usually are organized under the patronage of the Bulgarian President or the Prime Minister.

Willingness to bring together the representatives of all media that write or speak in Bulgarian gave rise to the idea of world meetings of the Bulgarian-language media. The purpose is to create networks of correspondents and contacts, which would encourage the development of professional interaction between the journalists. A unifying task of these forums is to make the Bulgarian media abroad work for safeguarding the Bulgarian identity and to help the integration of Bulgarians in a globalizing world. The expected outcome is "to make it so that the Bulgarians, irrespective if they are in the nation-country, or from the Diaspora, would know that there is a thing that unites them and this is Bulgaria" (Minchev 2005a).

In the course of the implementation of this approach, the media generally have displayed a surprising consensus. For the first Meeting in Sofia, out of the 108 known coordinates of the Bulgarian-language media across the world, 60 have been selected as meeting the following criteria: to be popular newspapers, magazines, TV or radio broadcasts; to wield influence in the Bulgarian communities; to maintain periodical publication; to have steady or high circulation (Minchev 2005b). All of the 60 invited Bulgarian media, from 25 countries, participated in the meeting.

The world meetings of the Bulgarian-language media are an unprecedented event in the history of Bulgarian journalism and media studies.

In a century of information technologies and boundless opportunities for communication, these forums mark the outset of a public discussion on the policies aimed at the Bulgarians abroad, as well as on the existence of the Bulgarian-language media abroad. By 2010, there had been six World Meetings of the Bulgarian-Language Media, and the main problems they tackled corresponded with some of the most pressing issues for the Bulgarian public and culture, as follows:

1. *The Bulgarian Media and Bulgaria's European Commitments (2005, Sofia, Bulgaria).* The object of discussion was the commitment of the media to informing the public about the pan-European process of enlargement and the challenges to national identity within the framework of united Europe (World Meeting 2005).
2. *The National Identity and the Role of the Media in Bulgaria's Image-Making (2006, Chicago, U.S.).* Topics concerning the professional standards in media performance were broadly discussed, as well as the positive and negative contribution of the Bulgarian-language media, at home and abroad, to Bulgaria's image-making across the world. Possible cooperation between participating parties was broached (World Meeting 2006).
3. *The Cyrillic Alphabet in Europe: The European Mission of the Bulgarian Media. (2007, Rome, Italy).* The meeting was held after Bulgaria's accession to the European Union. It was dedicated to the 1150th anniversary of the creation of the Cyrillic alphabet, one of the official alphabets in the European Union (World Meeting 2007).
4. *Media: The Obligatory "Third Party" in the Intercultural Dialogue (2008, Madrid, Spain).* The emphasis was laid on the issues related to the Bulgarian-language media across the world; the Bulgarian-Spanish cooperation in the field of media, culture, tourism, economy and politics; and presentation of the Bulgarian culture, language and history in the world (World Meeting 2008).
5. *Where To in the Real, Virtual and Informational Space? Tourism, High Technologies and the Media: The Bulgarian Arguments Against the Crisis (2009, Varna and Bourgas, Bulgaria).* Within the framework of this event a forum on tourism was held in Bourgas, as well as a business forum on high technologies and media in Varna. The aim was to use communication and interaction between the representatives of the media and business sectors with the greatest contribution to the GDP of the country in 2008 so as to discuss the prospects and to suggest constructive avenues for resolving some topical issues in both spheres (World Meeting 2009).
6. *The New East–West (2010, Vienna, Austria).* The forum focused on the evolution of the concepts of East and West and on reformulation of the relationships between them in the economic, social and cultural plan, and in terms of the media (World Meeting 2010).

CHARACTERISTICS OF THE BULGARIAN-LANGUAGE MEDIA ABROAD: RESULTS FROM A SURVEY

The challenges for the Bulgarian-language media abroad are exceptionally difficult and diverse when considered in relation to the media functioning within the frontiers of the nation-state. The work of the media abroad is considered by their representatives as a kind of missionary work, as they are conveyors of the national identity, the Bulgarian language and Bulgarian culture. The media abroad have direct contact with their audience and directly participate in the life of the Diaspora and it is this in particular that makes their work so diversified and interesting.

In order to obtain better knowledge about the Bulgarian Diaspora, a survey supported by the Scientific Research Sector of Sofia University was conducted by us among the Bulgarian-language media abroad in 2008.[2] Sixty-seven media bodies participated in the survey, comprised of: thirty-three newspapers (49.26%); seven magazines (10.48%); nine radio stations (13.42%); eight TV stations (11.94%); eight online media (11.94%). Data from this survey showed that most (over 60%) of the Diaspora media belong to the print media—newspapers and magazines. Most of the printed media published outside Bulgaria are free-of-charge. While newspapers account for nearly 50%, the radio, TV and online media claim 12% each. The number of the consumers of each media varied from one thousand to two thousand persons.

The questionnaire included nineteen questions, organized in four themes that aimed to explore the capacity and function of this media group; some of the findings from these themes are now examined.

The New Technologies and Contacts with the Audience

Irrespective of the fact that the Bulgarian-language media abroad are predominantly printed, the information and communication technologies occupy an important place in the everyday work of the journalists working there. Eighty percent of the respondents say that they value the significance of the new technologies in their contacts with the audiences, while 20% determine their role as relative.

The responses to the question about the type of feedback from the consumers are more varied. Twenty percent of the respondents have no opinion. The remaining 80% mention more than one type of feedback: for 70% these are telephone calls; for 40%, the traditional postal services; e-mail messages are pointed to also by 40%; 10% have assigned space in the medium for consumers' opinions, and another 10% contact consumers face-to-face at events organized by the medium.

The prevailing (40%) assessment of journalists about the role of the Web-based communities in the functioning of the media they work in is that they serve as an information source for the media. Thirty percent think that the Web-based communities play a great role, 10% answer that they play no role in the work of the Bulgarian-language media abroad, and 20% have no opinion on the question. Evgeni Kaidamov, representative of the *BG BEN* newspaper published in Great Britain (http://www.bgben.co.uk/news.asp?th), made an interesting comment:

> The web-based communities abroad are on a very high level; unfortunately they are still on a club level in Bulgaria. This is probably due to the mutual assistance that takes place in these communities abroad, which leads to institutionalization of their forums. For instance, the Internet forum http://bghelp.co.uk is a BG BEN project. The site could exist completely independently of the publication, which is no obstacle for the newspaper to borrow some of its most important topics from the forum. (Kaidamov BG BEN, 2008)

"In fact I am not a journalist, but a programmer," says Stoyan Mechkarov, owner of an Internet site of the Bulgarian community in Ireland (www.bulgariaie.com):

> When I arrived in Ireland, there had been absolutely no communication between the Bulgarians [in Ireland] and, given that I have knowledge about Internet, I try to do something which would create some kind of contact between us. In the last year and a half this site has been visited by over 20 thousand persons. Ninety-nine percent of these are Bulgarians and the rest are companies seeking advertising. The site is open and anyone who feels like it may write something in it. Even the Bulgarian Embassy approaches us to publish some announcements. The site is visited by many Bulgarians in Bulgaria who are interested in how to come and live in Ireland. (Mechkarov 2005)

These comments made by Mechkarov demonstrate that there are migrants who are attracted by the life in other countries about which they are able to learn more via the Internet-recorded experiences of their compatriots.

Specificities of the Diaspora: The Media-Diaspora Relationships

The assessments of the activity of the Bulgarian Diaspora vary from active to satisfactory, and some statements have even been heard that show the respondent's faith in their nation-country has been shaken, such as, "No one is interested in us." Some Bulgarian journalists abroad shared with us, during our survey, their personal opinions on this point:

Borislav Nikolov, Publisher of the *Forum* newspaper, Canada (http://www.forumbulgare.ca):

> There are about 40,000 Bulgarians in Canada. Our Diaspora there is marginal. Being Bulgarian is its secondary trait. The cultural life

is quite strong: movies, theatres, etc. Many Bulgarians attend such events. The Bulgarians come together in the churches. There are also Bulgarian schools which, however, are very miserable.

Neli Karagyozova, Chief Editor of *Atinski Vesti* (Athens News) newspaper, Greece:

> The Bulgarian societies in Greece have become increasingly active in recent years. Many cultural events are organized: celebration of the national holidays, concerts of Bulgarian performers, theme-oriented meetings, various interest circles, and evenings dedicated to Bulgaria. They keep alive the spirit of the Bulgarians away from the nation-country and also help popularize our national traditions and culture among the Greek public presenting our nation-country at its best. Already two Sunday schools operate, where the children learn Bulgarian language, history and geography.

Evgeni Kaidamov, *BG BEN* newspaper, England (http://www.bgben. co.uk/news.asp?th):

> The Bulgarian Diaspora in England is not very active. Bulgarians are passive people. They are active in the forums, where they can curse one or another institution. They lack any communal consciousness. The Bulgarians prefer to keep the useful information to themselves.

Assessment of the World Meetings of the Bulgarian-Language Media

The next group of questions related to the assessment of the World Meetings of the Bulgarian-Language Media. Eighty percent of the respondents found the idea positive, 10% were neutral and 10% had no opinion. The assessment of the organization of these meetings was positive for 60% of the respondents, negative for 20% and 20% had no opinion; some of the comments on the organization of the meetings are illustrated here:

> "The forums target goals outside journalism, behind-the-scenes diplomacy. Some rudiments of the propaganda abroad have been noticed. The Bulgarian State should allocate more funds and resources."

> "The Bulgarian media abroad know best the problems of the Bulgarian Diaspora. The world meetings may provide a rostrum for a direct dialogue between them, on the one hand, as speakers for the moods of the Bulgarian communities abroad and the representatives of the Bulgarian institutions, and on the other, as a transmitter of the problems of the émigrés and the Bulgarian emigration policy (the

ministries of social welfare, health and education, the State Agency for Bulgarians Abroad, etc.).”

“Lack of coordination. . .”

“There should be more contacts between the media and Bulgaria.”

”The most important thing is that we meet and exchange opinions. This has been realized. The rest is of secondary importance!”

The comments on the motivation to participate in the meetings mostly include professional interest:

“My motivation is both professional and personal.”

“Promotion of our media and getting acquainted with the problems of the Bulgarian communities across the world, with the trends in the Government policy related to the Diaspora; exchange of experience and an opportunity for a direct contact with colleagues from the same type of media.”

“Establishment of contacts with colleagues from different countries, familiarization with the problems and conditions of the Bulgarian media and Diasporas across the world.”

“More contacts between the media and Bulgaria in order to assist the motherland.”

The Bulgarian-language media abroad and the image of Bulgaria

The last group of questions targeted the potential of the Bulgarian-language media, both in influencing the Diasporas and in promoting the image of Bulgaria abroad. Only 20% replied that the Bulgarian-language media abroad can play a decisive role in the fulfillment of national causes. For 30%, the role of these media was relative, while 40% thought that there was no link between the Bulgarian-language media abroad and the Bulgarian national causes. Fifty percent of the respondents found that the Bulgarian-language media had a role to play in consolidating the Bulgarian national identity.

After the first meeting of the Bulgarian-language media in 2005, an association of the Bulgarian-language media was established. Sixty percent of the respondents regarded this idea as positive, 10% as negative and 30% had no opinion on the matter. The Association of the Bulgarian-Language Media in the World was registered in 2005, but has not yet achieved anything of significance as it has not met the commitments it undertook, namely:

- To become a coordinating centre for implementing of common ideas and initiatives
- To become an administrative body for coordinating the interaction between the Bulgarian-language media and the nation-state, which could be of help to the Diaspora
- To help promoting the achievements of Bulgaria and of the Bulgarians across the world
- To provide the Bulgarian-language media with information about the home country

This group of questions also included a query about the future of the world meetings of the Bulgarian-language media. Some of the possible topics for discussion refer to the issues concerning the second-generation emigrants, as well as to the migration policies of the Bulgarian government. The mass migration wave after the political changes of 1989 caught the country unprepared to cope with the disrupted balance caused by that migration; this is a key reason why one of the desired topics for future discussions is how the potential of successful Bulgarians abroad could be harnessed for the prosperity of the nation-country. Another proposed topic is connected with the professional standards, both in the Bulgarian-language media abroad and in the home country.

CONCLUDING WORDS

We live in times when borders and national affiliation tend to be regarded as conditional rather than as determinant and fundamental; under these conditions it is difficult to speak of national policies related to our compatriots abroad. However, one should not disregard the fact that for small countries such as Bulgaria, the contemporary world developments in migration are particularly painful, owing to the shrinking of the nation, both quantitatively and qualitatively. The topic of the nation-country is particularly sensitive to the Bulgarian Diaspora, but perhaps these Bulgarian communities abroad could be in some way legitimized with the help of the Bulgarian-language media. The rapid progress of the information and communication technologies has brought to the fore the issue of their impact on the global/regional/national/local communications environment. In the hypermodern times of the new millennium, humans exist both in real and in virtual environments. Under "glocalization"—dissolving of the global in the local—this is particularly obvious in the Diaspora communities and is a consequence of the new media realities.

It is clear from the findings of the survey that the Bulgarian-language media abroad have an enormous resource at their disposal which could be used to assist the Diaspora in maintaining their connection with their home nation. If these communities remain unaware of this opportunity and fail to use this resource, these media *de facto* cannot serve as a bridge between the nation of origin and the Diaspora. Nevertheless, as a result of the World Meetings of the Bulgarian-Language Media the Bulgarian citizens (living abroad and in Bulgaria) have been able to enhance their knowledge of the Bulgarian-language media abroad and of the problems of the Bulgarian communities there. Through these meetings mutual and beneficial professional relationships have now been established between the nation-country media and the Bulgarian media of the Diaspora. Indeed, what the six World Meetings of the Bulgarian-Language Media have managed to create is the feeling of a professional and national communion among the Bulgarian-language media abroad. Through these forums, and the use of online media, the urgent problems of the Bulgarian Diaspora have become public and the image of Bulgaria as an E.U. member-state has a new platform. The results of the survey affirmed that these results hardly would have been so encouraging if they had not taken place in the new broadband environment. These meetings have thus played a substantial role in the further building of the national politics towards the Diaspora.

NOTES

1. These NSI data on external migration from Bulgaria comprise information only on the persons who have officially reported a change in their address from the nation-country abroad. For comparison: the number of migrant Bulgarians in 2007 was 2,958, in 2008 it was 2,112 and in 2009, 19,039. http://nsi.bg/otrasal.php?otr=19&a1=367&a2=374#cont
2. Quotes included in this section were obtained from a survey conducted in Bulgaria by the authors during 2008.

REFERENCES

Agency for the Bulgarians Abroad. 2008. http://www.aba.government.bg/ (accessed October 10, 2008).

Appadurai, Arjun. 1996. *Modernity at large: The cultural dimensions of globalization.* Minneapolis: University of Minnesota Press.

Barbier-Bouvet, Jean-François. 2002. Internet, lecture et culture de flux. *Bulgaria Online.* http://www.online.bg/kultura/my_html/2230/Internet.htm (accessed October 10, 2008).

Constitution of the Republic of Bulgaria. 1991. National Assembly of the Republic of Bulgaria Web site. **http://www.parliament.bg/?page=const&lng=bg** (accessed October 10, 2008).

Dimitrov, Bojidar. 2010. I support the demand of the Diaspora to nominate its representatives at the Parliament, but the decision would belong only to the National Assembly. *EuroChicago.com.* http://www.eurochicago.com/2010/06/bojidar-dimitrov/ (accessed June 11, 2010).

Fidler, Roger. 1997. *Mediamorphosis: Understanding new media.* Thousand Oaks, CA: Pine Forge Press.

Friedman, Thomas. 2007. *The world is flat: A brief history of the twenty-first century.* New York: Picador.

Lash, Scott. 2004. *Critique of information.* Trans. into Bulgarian by Dimana Ilieva. Sofia: KOTA Publishing House.

Law for the Bulgarians Living Outside of the Republic of Bulgaria. 2000. http://www.careproject.eu/database/upload/BUotheroo1Trans.Eng.pdf (accessed October 10, 2008).

Mechkarov, Stoyan. 2005. Owner of an Internet site of the Bulgarian community in Ireland. *Politika (Sofia),* May 28–June 3, 16.

Meyrowitz, Joshua. 1985. *No sense of place: The impact of electronic media in social behaviour.* New York: Oxford University Press.

Minchev, Maxim. 2005a. For the first time we bring together our media from three continents. *Novinar Newspaper (Sofia),* April 7, 9.

———. 2005b. BTA brings together the Bulgarian media across the world. *Mission Newspaper,* May 19–25.

National Statistical Institute. 2011. Population Data. http://www.nsi.bg/otrasal.php?otr=19 (accessed April 30, 2011).

Todorov, Tzvetan. 1998. *In foreign land.* Sofia: Otvoreno Obshtestvo Publishers.

Toffler, Alvin. 1991. *The Third Wave.* Trans. into Bulgarian by Margarita Boeva. Sofia: Peyo K. Yavorov Publishers.

World Meeting of the Bulgarian-Language Media (1st). 2005. http://www.bta.bg/site/bg/initiative/initiativebta.htm (accessed October 10, 2008).

World Meeting of the Bulgarian-Language Media (2nd). 2006. http://www.bta.bg/site/bg/initiative2006/initiativebta2006.htm (accessed October 10, 2008).

World Meeting of the Bulgarian-Language Media (3rd). 2007. http://www.bta.bg/site/bg/initiative2007/initiativebta2007.htm (accessed October 10, 2008).

World Meeting of the Bulgarian-Language Media (6th). 2008. http://initiative2010.bta.bg (accessed June 10, 2010).

World Meeting of the Bulgarian-Language Media (5th). 2009. http://www.bta.bg/site/bg/initiative2009/initiativebta2009.htm (accessed October 10, 2009).

World Meeting of the Bulgarian-Language Media (4th). 2008. http://www.bta.bg/site/bg/initiative2008/initiativebta2008.htm (accessed October 10, 2008).

Religion, Mobility and Social Policies

How Migrants' Use of the New Media Is Shaping Society

10 "God is Technology"
Mediating the Sacred in the Congolese Disapora

David Garbin and Manuel A. Vásquez

INTRODUCTION

> When a revelation is coming, it's like a phone. A message comes to you. The ring is right here, in my heart [pointing to his heart]. Drii-innggg, driiingg. And then, I'll be stuck. Then I hear the voice talking to me. And it tells me 'do this, because I want this to be done right now!' God is holy, you know, God is technology. Everything we have right here, it's God's, thanks to God's spiritual existence. When you have a good relationship with God, the frequency is very good, the signal is strong. If you are good with your prayers, the communication is good. If you are messed up and if you are in a conflict with people, the reception is bad
>
> —Pastor Joshua, Congolese Pentecostal pastor, Atlanta, U.S.[1]

Pastor Joshua's experience of divine revelation reflects the importance of mediation but also suggests the absence of contradiction between technology and the realm of religious identities and practices. The example he quotes demonstrates that "old-time" religion is very much at home in the world of computer-mediated communications, and is far removed from the idea that religion might increasingly recede from the public sphere and become privatized with secularization and the inexorable advance of scientific rationalization. In fact, there might be a close "elective affinity" between modern forms of communication and a "glocal" Pentecostal performance of the sacred. When Pastor Joshua tells us that revelation is like a phone signal ringing in the believer's heart, he not only illustrates the strongly localized and embodied dimensions of the charismata, the gifts of the Holy Spirit, which are vital to the Pentecostal personal encounter with the sacred; in addition, the poignant expression "God is technology" and the metaphor of the electronic signal to represent divine calling points to the way in which religious transcendence dovetails with the dialectic of deterritorialization and reterritorialization generated by ICT networks. The simultaneously globalizing and localizing quotidian operation, computer-mediated communications mirror the powerful, anti-structural but deeply personal and immediate experience of the Holy Spirit. This mirroring facilitates the portability of the tight spirit-matter nexus in Pentecostalism: the idea that the Holy Spirit and the spirits serving the devil are engaged in a cosmic battle that is manifested in daily life across the globe. ICT networks, then, become a medium to transmit this worldview and a key tool to fight the cosmic battle on God's side, more specifically to conquer territory for Jesus.

The use of ICT networks as "sacralizing tools" is arguably most evident in the so-called Neo-Pentecostal churches, which explicitly link the experience of the sacred with material prosperity and health. Neo-Pentecostal theology holds that the world is a stage for a spiritual warfare between God and the devil. In this struggle, the task is to liberate Christians from spiritual bondage in order for them to lead healthy and successful lives. Liberation often takes the dramatic and dramatized form of exorcisms of demonic spirits in which the whole congregation performs a series of rituals that sacralize space and reconstitute the afflicted body. The spectacular events can be beamed, globally assisted by electronic media like the Internet, radio and television.

Pastor Joshua's example also highlights the important role that ICTs play in linking diasporic experiences and identity with religious performance. When migrants cross national borders, their religious identities, practices, and artifacts "accompany" them—along with "traveling spirits" (Hüwelmeier and Krause 2010). The study of transnational migration and global diasporas has gradually recognized the often central role that religion plays in the multiscalar formation of collective and individual identities. Furthermore, while the circulation of worshippers, missionaries, religious experts and entrepreneurs is an important component of religious transnational fields, the physical mobility of individuals and groups is not the only pre-requisite for the production of diasporic (sacred) communities, as Peggy Levitt (2007) and others have argued. Thus "crossing the electronic frontier" is a central element of the "globalization of the sacred" (Vásquez and Marquardt 2003), as spiritual power is increasingly transmitted through images and sounds that, due to rapid changes in communication technology, circulate at blinding speeds, becoming localized through the interplay with "autochthonous" narratives and practices. This transmission through particular imagescapes has contributed to the production of new transnational charismatic fields of religious performance and representation (Meyer and Moors 2006; De Witte 2003).

In this chapter, we explore the creative use of electronic media to advance charismatic and prophetic/messianic Christianity among migrant and diasporic populations located at various nodes of an asymmetrical power-geometry (Massey 1993). What is the nature and function of ICT networks between places in the diaspora and places in the homeland, and between places *across* the diaspora? What can transnational communication strategies and practices tell us about the symbolic and sacred importance of social spaces within a global diasporic territory? To which extent is the use of media connected to the production of religious identities and embedded in particular politics of the sacred?

We take African religious scapes as a case study, drawing on ongoing fieldwork in the Congolese diaspora in London and to a lesser extent in Atlanta and the Democratic Republic of the Congo (DRC).[2] We will show how ICTs are deployed by Congolese religious actors to draw and redraw symbolic geographies of the sacred as they negotiate the multiple (dis) embeddedness that accompany diasporic and migrant livelihoods. We will first examine the role of ICT mediation in the performance of the sacred and in the diasporic configurations of Pentecostal/charismatic churches. In our second case study, we will focus more specifically on the Kimbanguist church, one of the largest transnational Congolese churches, in order to analyze the linkages between diasporic belongings, politics of the sacred and ICT mediation.

"TAKING TERRITORIES": CONGOLESE PENTECOSTALISM, TRANSNATIONAL COMMUNICATION AND DIASPORIC NETWORKS

Perhaps the most striking feature of the Congolese religious sphere in the DRC and the diaspora is the dominance of Pentecostal churches. One of the expressions used to designate the Pentecostal/charismatic churches, *Eglises de Réveil*, refers to a religious revival and awakening (*réveil*) which dramatically altered the religious landscape of the DR Congo. The Pentecostal *réveil* in the Congo, which emerged at the end of the Mobutist era against the backdrop of increasing violence, political instability and endemic economic crisis, has not only impacted on the religious domain. As Katrien Pype (2006a) has shown, Pentecostalism has introduced a new "charismatic habitus" (Coleman 2000), a new embodied aesthetics and praxis manifested in urban dances, popular culture and other public performances in Kinshasa. She describes how the public culture sphere has become increasingly charismatic with, for example, the conversion of hundreds of "worldly" music bands and theatre groups to "purified" evangelical Christian forms of performing arts. In addition, the "key scenario" of charismatic Christianity, the fight between godly forces over evil forces, the spiritual warfare against the world of *féticheurs* and *magiciens*, has been established as a dominant plot—for instance in the popular *maboke* (theater/TV dramas)—with, at its core, *pasteurs* and *prophètes* as "cultural heroes" of the new moral economy of the city (Pype 2009).

Migration has been a key factor in the emergence of a Congolese religious diasporic landscape dominated by Pentecostal churches. This diasporic landscape has diversified with the recent migration of Congolese outside francophone countries, such as Germany, Holland, the U.S., Canada or the U.K. (Garbin and Pambu 2009). For a Congolese church created abroad, keeping or setting up a branch in the homeland can enhance its legitimacy, its "Congoliness", but also its spiritual identity. When Pastor Joshua talked about the strategy for his church, he described a symbolic and spiritual geography with, at its heart, a particular complementary logic between the homeland and diaspora. Thus, he mentioned how the development of the Kinshasa branch of his church, which he is managing and financing from Atlanta, could provide a "spiritual power" superior to what could be gained from his church in the U.S., as it was easier to mobilize a large number of worshippers in Kinshasa. Through this kind of religious or spiritual "outsourcing", worshippers in Kinshasa provide the power of prayers, the "spiritual engine", while the U.S. branch provides the financial support required for the material development of the church in the Congo.

The visit of pastors from the DRC is an important element of the logic of continuity between homeland and diaspora, even if it is inscribed within a wider, uneven—yet complementary—translocal geography. However, these visits, sometimes called "crusades" in the case of the most renowned pastors, often prove to be difficult to organize, as visas to the Europe or to the U.S. are not easy to obtain. In addition, the activities of the radical opposition in Britain, *les combattants*, has had some impact on the circulation of religious leaders—but also artists and musicians—criticized for supporting the current regime and threatened with violence if they come to the U.K. To overcome these limitations, congregations take advantage of transnational media networks, Internet and satellite TV but also inexpensive and easily accessible video technologies, which allow them to record on the highly portable and durable DVD format various events including Sunday services, sermons, religious conferences or certain rituals (such as collective baptisms, or "deliverances"). This circulation of DVDs not only enables the transmission and re-enactment of "religious effervescence" between congregations across translocal diasporic spheres, it also contributes to the embodied Pentecostal aesthetic that Pype (2009) has documented. Here, religious DVDs operate as a privileged medium for "social remittances", that is, for "ideas, behaviors, and social capital that flow between receiving and sending communities. They are tools with which ordinary individuals create global culture in a local context" (Levitt 2001, 11).

Another transnational configuration in the religious sphere concerns the diasporic development of churches based in the Congo, mainly through the process of migration, which is often reinterpreted in terms of a "reverse-mission", part of a divine plan—the "return of the Gospel"—to bring moral and spiritual regeneration to the secular and disenchanted West (Mary 2002; Maskens 2008; Fancello 2006; Ter Haar 1998; Ugba 2009). In that scenario, the church in the DRC often relies on the diaspora for financial and symbolic support. Indeed, while setting up a Congolese branch can enhance the "authenticity" of a U.K.-based Congolese church, the extension of a DRC-based "parish" to the diaspora is a great source of prestige. For a Congolese church, having a foothold in the diaspora not only represents the fulfillment of the "Great Commission" through which Jesus enjoins the Apostles to go make "disciples of all nations", but also forges a connection to modernity and globalization, expanding the scope of a "health and wealth" gospel. The competition is fierce in the charismatic sphere of the DRC and the religious mediascape (of Kinshasa in particular) is a privileged site of struggle for recognition between churches (Pype 2006b). Video or audio recordings from diasporic branches can be broadcast in the Congo via a plethora of religious TV channels or radio stations. These broadcasts represent precious symbolic capital for *pasteurs* and *prophètes* engaged in the "spiritual crusade" against evil forces and who are also keen on strengthening their legitimacy.

"Mega-churches" are also relying on transnational ICT flows and networks that link the local, national and global. For instance the Kingsway International Christian Centre (KICC), a Pentecostal church created by a Nigerian pastor in London in 1992, tends to attract a more educated and more upwardly mobile section of the Congolese community. One of the mottos of the church, "*Raising champions, taking territories,*" evokes the idioms of expansion, evangelization and conquest, which are central in the discursive construction of a charismatic collective identity. For KICC, this charismatic identity is closely bound up with the appropriation of modern media, as one could read in the presentation of the media ministry on the KICC Web site:

> KICC's goal is to build a church "without walls", and this can be achieved through our media ministry—Winning Ways. Winning Ways is a major KICC outreach programme that encompasses both print and electronic media, including radio and television. God has appointed KICC with a holy mandate to reach a dying world with the Living Word. Winning Ways is now aired in 25 stations around the world, through a number of these key partnerships, particularly with God TV (which alone reaches 217 nations) and TBN Africa (which reaches 13 countries in the southern part of the continent). We are also aired on a number of local stations around the world. We have already exceeded our target of being in 200 nations by 2010. This means we are touching the world for Jesus, bringing a message of hope, healing, salvation and deliverance. Our media strategy is clear and focused. We are continuing to form strategic alliances with media partners across the globe that will enable us to be effectively used by God to carry the good news to His people.[3]

Thus for KICC, and many other large Pentecostal churches, the idea of a "church without walls", a strategy embedded in a dialectic of territorialization/deterritorialization, is a powerful way to convey and mediate the message of God across national borders. In the next case study, we will explore how this mediation of the divine and the sacred operates in the

transnational space of a large Congolese Christian movement, the Kimbanguist church.

THE KIMBANGUIST CHURCH: FROM LOCAL
PROPHETIC MOVEMENT TO RELIGIOUS DIASPORA

Regarding the ICT use in the construction of a transnational and imagined community of believers, the Kimbanguist church relies on a different kind of "ideal" diasporic configuration with a strong holy center,
Nkamba, located in the Democratic Republic of the Congo. Nkamba
is where Kimbanguism, a Christian prophetic movement led by Simon
Kimbangu, emerged, initially among the Bakongo in the colonial context
of the then-Belgian Congo. Representing a threat to the colonial order,
Kimbangu was arrested by the Belgian army in 1921 only a few months
after the start of his ministry of faith healing, which attracted thousands
of pilgrims to Nkamba (Sarró and Blanes 2009). This spiritual revival
triggered the emergence of other local *ngunza* (prophets) inspired by the
divine and miraculous character of this "modern Kongo prophet" (Mac-
Gaffey 1983). After thirty years of imprisonment, Simon Kimbangu died
in jail in 1951 in Elisabethville (now Lubumbashi), eight years before the
official recognition of the church by the Belgian authority. In the post-
independence and Mobutu years the Kimbanguist church grew to become
a major institution in the Congo and now has a strong presence in central
Africa and across the Congolese diaspora, especially in Europe.

To understand the role of transnational and ICT connections within
the Congolese Kimbanguist diaspora one needs to take into account the
symbolic, spiritual and political importance of Nkamba. Kimbanguist
identity articulates several "diasporic horizons" (Johnson 2007) revolving
around pan-African and Afro-centric tropes of collective migration; exile
and "return" to Africa and Nkamba has become the center of an imagined and prophetic transnational "Black Atlantic". Nkamba thus occupies
a pivotal place in the spiritual life of Kimbanguists, perhaps even more so
in the diasporic context. The development of Nkamba, the New Jerusalem—whose sacred geography has been shaped by the presence of multiple
biblical "signs" (*bilembo*)—is thus seen as a duty, a central part of the holy
work (*misala*) for all Kimbanguists.

In addition to these post-colonial, pan-African and biblical dimensions,
for Kimbanguists in the diaspora the powerful role of Nkamba is closely
linked to the presence of the current leader of the church: "Papa" Simon
Kimbangu Kiangiani, a grandson of Simon Kimbangu, represents a vital
"sacred linkage". While most Kimbanguists have their "portable" version
of Nkamba, the earth and above all the holy water,[4] this direct connection
with the spiritual leader is perceived as essential. Thus he is regularly contacted by Kimbanguists from the diaspora requesting a prayer or blessings,

often for healing purposes, divination or advice before an important decision or to solve a problem (related to work, marriage or immigration status). To speak to *Papa* one has first to contact a member of *le protocole*, who acts as gatekeeper and who, sometimes after negotiation, will authorize the communication, through mobile phone, with the spiritual leader. In that sense, this "sacred communication", this link with the spiritual authority, is often embedded in particular politics of access, gatekeeping and networking, and, according to our own observations, calls from the diaspora tend to be treated as a priority.

SACRED MEDIATION AND THE TRANSNATIONAL CENTER-PERIPHERY NEXUS

Like many religious groups in the Congo, the Kimbanguist church has its own TV and radio stations. These are managed by the RATELKI (Radio Télévision Kimbanguiste) and while they primarily broadcast in the DRC, worshippers in the diaspora can access the Kimbanguist audio-visual world through digital technology. Most of the DVDs circulating in the diaspora have been produced in Europe, particularly in Belgium or France, and through this control of the technology linked to the performance and mediation of a Kimbanguist global identity, the diaspora tends to become gradually less peripheral in relation to the church in the Congo (Mélice 2006). While this suggests a process of decentralization and polycentralization at work within the Kimbanguist public sphere, it is also true that the contents of this digital mediation (messages from the spiritual leader and recordings from important festivals held in Nkamba) often tend to reinforce both the legitimacy of the spiritual leader and the "centering" of Nkamba. This reaffirmation of a strong spiritual and institutional center is especially important given the existence of both internal schismatic divisions (linked to the inheritance of the leadership of the church) and external criticisms of the evolution of a Kimbanguist "Christian" theology.

In terms of the use and role of electronic media, while (uneven) access to ICTs is developing in Kinshasa, Nkamba currently has, at the time of writing, no stable Internet connection. When asked about this electronic (and physical) isolation of Nkamba—an apparent paradox, given the renewed "centering" of Nkamba[5]—Kimbanguists in the diaspora sometimes point out that access to the divine requires sacrifice and true faith. In other words, there is no highway to heaven, a salient narrative in times of pilgrimages when thousands travel, often for days, to the remote Holy City. Yet some in the diaspora are also advocating change and the modernization of Nkamba (especially its electrification), arguing that Kimbanguists abroad, with their skills and resources, have a strategic role to play in this development. Many also refer to the "resistance" of negative and evil forces, whether spiritual or humans, blocking the development of the Holy City. This resistance, these *freins spirituels*, will eventually be overcome, many believe, by a power-ful divine intervention instituting Kimbanguism as a true "hope of the world".[6] In that sense, the (re)connection of Nkamba as a spiritual uni-versal center not only constructs a specific symbolic geography of the sacred but is also integral to a wider prophetic and millenarist temporal-ity, the promises (*bilaka*) of a sanctified future when Nkamba, the "New Jerusalem" in Africa, will "welcome all nations" and will be the place from which global salvation and redemption originates.One of the main implications of the current "electronic isolation" of Nkamba is that most of the Kimbanguist Web sites are hosted in the diaspora. In addition to textual material and official literature of the church, these Kimbanguist Web sites contain videos (also posted on YouTube) of celebrations, fes-tivals and sermons recorded in Nkamba. This allows the circulation of sacred images, words and sounds and facilitates the interconnection of "virtual territories" within the globalized Kimbanguist public sphere. Moreover, this mediation is important insofar as it creates a political/ organizational link between the sacred centre, the homeland and the diaspora. This transnational link, in particular the connection with the spiritual leader in Nkamba, is perceived to be vital for the reinforcement of the faith of those in the diaspora—especially the youth, who may not have firsthand experience of the Holy City or know much about the his-tory of Kimbanguism:

> Thanks to the Internet people can now listen to the messages of Papa [the spiritual leader] in Nkamba. For example, on Tuesday there is always a service especially for Kimbanguist youth, with Papa Mbenza [brother of the spiritual leader]. The leader delivers his mes-sage which is broadcast live on Ratelki.com, from Belgium. They connect with Nkamba via mobile phone. So I listen to the message, I transcribe it and translate it in English, and print it as a leaflet for the youth. It's a good thing for our youth, it reduces the distance with Nkamba and I think it's useful to reinforce their faith. (Interview with Kimbanguist youth leader, London)

A recent event held in the diaspora, the inauguration of a new place of worship in London, also illustrates the importance of bridging the distance with Nkamba through transnational communication. At the end of the celebrations, Kimbanguists managed to reach the spiritual leader's mobile phone in the Congo through Skype-related technology. In fact, this con-figuration reflected a technological *bricolage* as the Internet access was obtained through an unsecured wireless connection with an irregular and rather weak signal. Nevertheless, all the worshippers heard the voice of the spiritual leader once the connection between the laptop computer and the sound system had been established. An intense feeling of joy could be perceived among the churchgoers when the leader congratulated, from

Nkamba, the members of the London parish, gave his blessings and prayed for the success of this new *paroisse* in the U.K.

Electronic media can also, however, define the boundaries of transnational cyberspaces that are disconnected from the homeland and the Holy City. This is the case for Facebook communities recently formed by Kimbanguist youth living in different European countries. Through this virtual networking, the hybrid modes of belonging of these youth are played out, and social relations are forged or maintained across national boundaries. These connections are sometimes activated or reactivated "offline" during trips across Europe for weddings, funerals or important Kimbanguist festivals, but within this more autonomous public (cyber)sphere youth can evade the gaze and social control of adults and elders. Increasingly more youth seem to consider ICTs as a medium to make Kimbanguism more recognized and visible in the "host" context, in Europe. One of the traditional ways for the church to gain public visibility is through its brass band (Fanfare Kimbanguiste, or FAKI).[7] The FAKI, which performs divinely "inspired" hymns,[8] plays a crucial role during religious rituals and also regularly performs for a wide range of events in the Congo. In the diasporic context the brass band, composed mainly of youth, takes part in community events (Congolese funerals, weddings) but also increasingly in public events,[9] the most prominent being London's New Year Parade, during which the urban center of the global city became a stage for the performance of a Kimbanguist identity. In addition to this temporary appropriation and sacralization of space (Garbin 2010c), FAKI members were also keen to underline how the visibility and evangelizing mission linked to the parade could be achieved through global media—via TV broadcast[10] or through video clips on YouTube:

> The [New Year] parade increased the visibility of the church . . . and the fact that it went on YouTube and the comments that people left. It was very important for us to reach that audience after being here [in the U.K.] for such a long time. And in the essence of Kimbanguism, it exists to be spread, it's the message of God. We can't spread it if we just stick to a small place in Docklands or in Edmonton [locations of London Kimbanguist churches] . . . We have never been on that stage before. We stood up and we did it! (Interview with member of the brass band)

While these enactments and urban performances of identities move beyond the local, as images and sounds are circulated through ICTs, the role of the diasporic periphery in the diffusion and mediation of the sacred is also increasingly debated. One of the most popular productions originating from the diaspora, but also one of the most discussed among Kimbanguists, is a program recorded and distributed on DVD called "Emission Ya Bazoba". Containing news about the churches, videos of celebrations, messages from the spiritual leader or sequences about Kimbanguist history

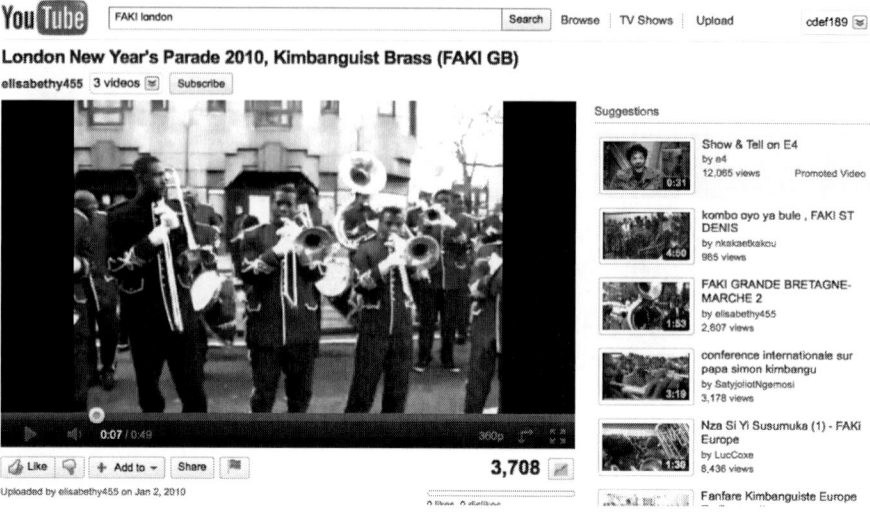

Figure 10.1   Video uploaded on YouTube by member of the brass band.

and prophecies, it sometimes additionally features interviews with non-Kimbanguists. Among them are religious leaders, preachers, healers or *ngunza*—(mostly Bakongo) "prophets"—who claim to be inspired by Kimbangu or acknowledge his legacy in their own religious and spiritual work. While for the producer of Emission Ya Bazoba, the aim of these interventions is to draw attention to the influence and significance of a Kimbanguist tradition in prophetic (or neo-prophetic) movements, many have criticized what they consider a tendency to confuse "official Kimbanguism" with other Bakongo prophetic traditions or politico-religious movements (Garbin 2010a, 2010b).

Parallel to these critiques there is also a growing concern that through blogs, Web sites, discussion forums or DVDs created by non-Kimbanguists, the "message of Papa" may be distorted and the theology of the church misrepresented. Debates taking place in the Congo or in the diaspora, for instance on the Christian or Kongo authenticity of the Kimbanguist church or about its role during Mobutu's or the current regime, have found an echo and are amplified on the Internet or diasporic African satellite TV networks. This, in turn, leads more and more Kimbanguists to consider the mediascape of ICTs as an important territory to appropriate, not only to evangelize but also to draw clear-cut boundaries between the Kimbanguist church and other prophetic traditions or to respond to critiques about the theology or orientation of their church.

The complex dynamics of center-periphery linked to the "routinization" and progressive diasporization of a prophetic movement such as Kimbanguism is not, of course, solely contingent on processes of mediation through ICTs (Eade and Garbin 2007; Mélice 2001, 2009). Indeed, the role and functions of the ICT mediation within this framework may change, according to the evolution of organizational processes, the development of new bureaucratic infrastructures or the modernization of Nkamba. Additionally, there is also an ambivalent dimension in the ways in which the use and status of media and ICT is interpreted by religious leaders in diaspora—and not only within the Kimbanguist church. Internet, satellite TV channels, YouTube, Facebook or chat rooms, are often presented as "distraction" or even as powerful media through which devil and occult forces can operate. Yet they are also increasingly useful in order to adapt the "message" to youth or simply to reach them:

> Facebook or MSN are distractions for the youth. This world of entertainment can control them, it has strong, negative forces. But we also need to use it to serve the church. Youth don't answer my emails but they do use Facebook, so I use it all the time. (Interview with Kimbanguist youth leader)

This comment echoes a sermon heard in the French-speaking branch of KICC, in London, when the pastor urged the faithful to "move from the Ipod to the Wepod." By this, he was deploring the individualized tendency of new media use while suggesting their potential strategic role for strengthening the cohesion and collective identity of the "brotherhood in Christ".

CONCLUSION: TRANSNATIONALISM, RELIGIOUS MEDIATIZATION AND THE IM-MEDIACY OF OHE SACRED

Beyond physical mobility, through migration or mission, a widespread process of mediatization has also enabled the contemporary globalization of religion (Csordas, 2007). Religious mobility is thus not limited to the physical circulation or migration of worshippers, religious leaders or experts, but also involves the transnational flows and networks of capital, images, sounds and symbols. DVDs, the Internet, radio or TV broadcasts have become important instruments for the expansion of "sacred territories" and in particular for Pentecostal churches. These tools have also the potential to provide a powerful embodied and emplaced sense of imagined community, "sodalities of emotion", as Appadurai (1996) calls them, linking believers across national borders. Under these conditions, the Internet "is not primarily or essentially about the 'virtualization of the real'" (Vásquez and Marquardt 2003, 118) obliterating traditional spatio-temporal scales to produce a placeless cyberspace. Rather, the "Internet represents a space where the faithful can reinscribe the local, where the intimacy and immediacy of the physical [center] can be stretched, beamed globally to and experienced vicariously in diaspora" (Vásquez and Marquardt 2003, 118). Thus, the actual performance of the gifts of the Holy Spirit in Pentecostalism and the purifying and heal-ing prophetic power of the sacred axis mundi in Kimbanguism become portable through ICTs, allowing dispersed communities throughout the world to partake of the pneumatic power of personal embodiment and the holy center. However, as suggested by our case study of the Kimban-guist church, it is also important to recognize the differential access to new communication technologies and the (diasporic) power-geometries involved by the uneven geography of ICTs (Latham and Sassen 2005).

Moreover, the imbrications of ICTs and charismatic Christianity have important implications for young people in diaspora, as they seek to nego-tiate the demands and expectations of their parents and the challenges of incorporation into "hyperdiverse", potentially "immoral" environments. Here the increasing use, particularly among Neo-Pentecostal churches like KICC or the Congolese *Eglises de Réveil*, of practices and media common in "profane" popular culture enables young people in diaspora to build alter-native, hybrid identities and spaces of belonging. A key dimension of these hybrid identities and spaces is the ability to consume the products of global modernity safely while building stable identities anchored in simple ethi-cal principles that neatly divide good from evil, sacred from profane, and authentic from inauthentic. In that sense, charismatic Christianity molds a redeemed body with a habitus more in line with the spirit of late capitalism.

The genius of pneumacentric Christianity has always been its portabil-ity, the capacity to mediate im-mediacy, to generate the same transforma-tive effervescence in ever new contexts, from small churches in the urban *parcelles* of Kinshasa to tent revivals and slick mega-churches in London. ICTs represent a new articulation of charismatic Christianity's age-old dialectic between immediacy and mediatization. The ability of the new electronic media to produce "hyper-reality" (Baudrillard 1994), to create deterritorialized virtualities that are more vivid and all-encompassing than mundane reality, lends itself to the instantaneous and simultaneous con-sumption of the sacrality across multiple sites and scales. In other words, ICTs allow for the transnational unbounding of the "raw power" of the Holy Spirit. When a Pentecostal pastor such as Pastor Joshua engages in power-prayer that is simulcast to other temples in the diaspora, them-selves local stages in the cosmic spiritual combat between good and evil, mediatization functions to magnify and project globally the immediacy of charismatic Christianity. Here the boundaries between mediatization and immediacy, between virtuality and reality, begin to blur as part of the process of (re)sacralization.

This is not to say that the pneumatic power unleashed by ICTs is always safe. As we saw in the case of the Kimbanguist church, the use of ICTs is embedded in particular politics of religious legitimacy and authenticity and may have unintended, and even unwanted consequences. The more mediation there is, the higher the danger of conflicts over orthodoxy and the greater the probability of emergence of alternative nodes of sacrality which may contest the authority and power of a sacred axis mundi like Nkamba. However, despite these dangers, African diasporic churches have come to rely heavily on electronic media and global popular culture to link their nodes and (virtual) sacred territories, but also to generate the collective spectacles that are the essential ingredient for the personal experience of the Holy Spirit, fusing the medium and message in ways that even Marshall MacLuhan would never have anticipated.

NOTES

1. From interview with David Garbin and Manuel Vásquez conducted in Atlanta, August 2008. The name of the Pastor has been changed to protect his anonymity.
2. As part of two research projects: *The Religious Lives of Migrant Minorities* (funded by the Ford Foundation through the Social Science Research Council, USA) and *Performing Identities and Spaces among Brazilian and Congolese Immigrants in London and Atlanta: The Case of Two Transnational Religious Networks* (funded by the American Academy of Religion).
3. http://www.kicc.org.uk/TVRadio/KICCAirforce/tabid/76/Default.aspx
4. Taken from the sacred pond used by Kimbangu to heal, and where the pilgrims bathe.
5. In 2010 all national leaders from Europe were asked to spend several weeks in Nkamba for "spiritual training", but also in order to reinforce and strengthen the ecclesiastical uniformity and bureaucratic centralization.
6. One of the main mottos of the church
7. The role of the brass band, created in the 1960s, is closely linked to the idea of an evangelical mission because its main aims were to attract new members and to announce the official recognition of the church.
8. The hymns also sung in Kimbanguist choirs are directly "inspired", received through dreams and visions by those who possess this charismatic gift.
9. E.g. Black History Month and the Hackney Carnival in London. In France, Kimbanguist musicians contribute to the nation-wide *fête de la musique*.
10. E.g. Sky News, CNN, BBC.

REFERENCES

Appadurai, Arjun. 1996. *Modernity at large: The cultural dimensions of globalization*. Minneapolis: University of Minnesota Press.

Baudrillard, Jean. 1994. *Simulacra and simulation*. Ann Arbor: University of Michigan Press.

Coleman, Simon. 2000. *The globalization of charismatic Christianity: Spreading the gospel of prosperity*. Cambridge: Cambridge University Press.

Csordas, Thomas J. 2007. Introduction: Modalities of transnational transcendence. *Anthropological Theory* 7(3):259–272.

De Witte, Marlene. 2003. Altar media's living world: Televised charismatic Christianity. *Ghana, Journal of Religion in Africa* 33(2):172–202.

Eade, John, and David Garbin. 2007. Reinterpreting the relationship between centre and periphery: Pilgrimage and sacred spatialisation among Polish and Congolese communities in Britain. *Mobilities* 2(3):413–424.

Fancello, Sandra. 2006. *Les aventuriers du Pentecôtisme Ghanéen: Nation, conversion et délivrance en Afrique de l'Ouest.* Paris: IRD-Karthala.

Garbin, David. 2010a. Embodied spirit(s) and charismatic power among Congolese migrants in London. In *Summoning the spirits: Possession and invocation in contemporary religion,* ed. Andrew Dawson, 40–57. New York: IB Tauris.

———. 2010b. Symbolic geographies of the sacred: Diasporic territorialisation and charismatic power in a transnational Congolese prophetic church. In *Traveling spirits: Migrants, markets and mobilities,* ed. Gertrude Hüwelmeier and Kristine Krause, 148–164. London: Routledge.

———. 2010c. Evangelisation as urban *dawa*: Moral subjects in (re-)moralised landscapes. Paper presented at the workshop *African Christianity in Britain: Citizenship and the Migrant Subject,* University of Sussex, Department of Anthropology, School of Global Studies, June 7.

Garbin, David, and Wa Gamoka Pambu. 2009. *Roots and routes: Congolese diaspora in multicultural Britain.* London: CRONEM/COREGOG.

Hüwelmeier, Gertrud, and Kristine Krause. 2010. *Travelling spirits: Migrants, markets and moralities.* London: Routledge.

Johnson, Paul C. 2007. *Diaspora conversions: Black Carib religion and the recovery of Africa.* Berkeley and Los Angeles: California University Press.

Latham, Robert, and Saskia Sassen, eds. 2005. *Digital formations: IT and new architectures in the global realm.* Princeton: Princeton University Press.

Levitt, Peggy. 2001. *Transnational villagers.* Berkeley: University of California Press.

———. 2007. *God needs no passport: Immigrants and the changing American religious landscape.* New York: The New Press.

MacGaffey, Wyatt. 1983. *Modern Kongo prophets: Religion in a plural society.* Bloomington: Indiana University Press.

Mary, André. 2002. Pilgrimage to Imeko (Nigeria): An African church in the time of the "global village". *International Journal of Urban and Regional Research* 26(1):106–120.

Maskens, Maïté. 2008. Migration et Pentecôtisme à Bruxelles. *Archives des Sciences Sociales des Religions* 143:49–69.

Massey, Doreen. 1993. Power-geometry and a progressive sense of place. In *Mapping the futures: Local cultures, global change,* ed. Jon Bird, Barry Curtis, Tim Putnam, George Robertson, and Lisa Tickner, 59–69. London and New York: Routledge.

Mélice, Anne. 2001. Le Kimbanguisme: Un millénarisme dynamique de la terre aux Cieux. *Royal Academy of Overseas Science, Bulletin des Séances* 47:35–54.

———. 2006. Un terrain fragmenté: Le Kimbanguisme et ses ramifications. *Civilisations* 49(1–2):67–76.

———. 2009. Le Kimbanguisme et le pouvoir en RDC: Entre apolitisme et conception théologico-politique. *Civilisations* 58(2):59–80.

Meyer, Birgit, and Annelies Moors. 2006. *Religion, media, and the public sphere.* Bloomington: Indiana University Press.

Mokoko, Gampiot. 2010. *Les Kimbanguistes en France.* Paris: L'Harmattan.

Pype, Katrien. 2006a. Dancing for God or the devil: Pentecostal discourse on popular dance in Kinshasa. *Journal of Religion in Africa* 36(3–4):296–318.

———. 2006b. I do not want to marry my ngatiul': Mass-mediated alliances between youngsters and Pentecostalism in Kinshasa. EASA e-seminar. http://www.philbu.net/media-anthropolgy/workingpapers.htm.

————. 2009. "We need to open up the country": Development and the Christian key scenario in the social space of Kinshasa's teleserials. *Journal of African Media Studies* 1(1):101–116.

Sarró, Ramon, and Ruy Llera Blanes. 2009. Prophetic diasporas: Moving religion across the Lusophone Atlantic. *African Diaspora* 2(1):52–72.

Ter Haar, Gerrie. 1998. *Halfway to paradise: African Christians in Europe*. Cardiff, Wales: Cardiff Academic Press.

Ugba, Abel. 2009. *Shades of belonging: African Pentecostals in twenty-first century Ireland*. Trenton, NJ and Asmara, Eritrea: Africa World Press.

Vásquez, Manuel A., and Maria Friedmann Marquardt. 2003. *Globalizing the sacred: Religion across the Americas*. New Brunswick, NJ: Rutgers University Press.

11 Mediatized Migrants
Media Cultures and Communicative Networking in the Diaspora

Andreas Hepp, Cigdem Bozdag and Laura Suna

INTRODUCTION: A DIFFERENT PERSPECTIVE ON MEDIA AND MIGRATION

In this chapter we want to develop a different perspective on media and migration. It is specifically not our intention to make any rash statements on the role of certain media in the "integration" of "ethnic minorities" into "national host societies" (see for this discussion Cottle 2000). Rather, we want to formulate some considerations on how we can capture, on the one hand, the multidimensionality of diasporic media cultures (an approach that is typical in present media ethnography) without forgetting, on the other hand, that there are typical patterns of media appropriation across migrant groups. The foundation for this is an empirical study on the media appropriation and communicative connectivity of the Moroccan, Russian and Turkish diaspora in Germany. Based on this study we have developed the concept of "mediatized migrants". This concept argues that we must understand the present culture of migrants as *media* cultures, because we are now only able to comprehend them in the context of media communication. In this sense, migrants are nowadays "mediatized"; that means that their articulation of a migrant identity is deeply interwoven with and molded by different forms of media. However, the diasporic media cultures of mediatized migrants remain highly differentiated, and they are marked by conflicts and contradictions. Our empirical research on the Moroccan, Russian and Turkish diaspora in Germany demonstrates that this multiplicity can be described across the different migrant communities alongside a typology of origin-, ethno- and world-oriented migrants.

FROM "CONNECTED" TO "MEDIATIZED MIGRANTS": APPROACHING DIASPORIC MEDIA CULTURES

Some years ago Dayan (1999) emphasized that it was not appropriate to reduce the research field of media and migration only to the issue of integrating migrants through mass media in a hosting national society. In his reflections on the status of media for building a diasporic community, he argued that besides a "big" mass media, "small" media of personal communications—like the telephone, letters, family videos, etc.—are especially important for the cohesion of diasporas. Almost ten years later, Bailey, Georgiou and Harindranath (2007, 2) used the concept of "diasporic media cultures" to describe just such a mediated cohesion of diasporas. However, in their view, these media cultures are not so much the media cultures of the "small" media, but much more the media cultures constituted by the various offers of mass media addressing different members of a diaspora.

A stronger focus on digital media (the Internet, mobile telephone) is shared by Diminescu when she speaks of the "connected migrant" (2008). Via this concept she suggests that the advent of the Internet and mobile telephones introduced a comprehensive communicative connectivity into migrants' lives. While they stay on the move, their lives are marked by the accessibility of VOIP (voice over Internet), e-mail, music downloads from their place or country of origin, etc.: "Yesterday the motto was: immigrate and cut your roots; today it would be: circulate and keep in touch" (Diminescu 2008, 568). In times of globalization the potential of digital media molds the everyday lives of nearly all people living in Western societies (Tomlinson 1999, 150–180). However, these media have a special capacity for migrants, as via their use the quality of diasporic culture changes. Also, when living abroad and being dispersed, migrants are now far less separated from each other communicatively. In contrast, they have various possibilities for maintaining their (previous) communicative relationships with their place or country of origin or for building up (new) communicative relations in the diasporas and their present living context. Consequently, one can state that the (media) culture of "connected migrants" is additional to the previous culture of migrants.

Diminescu captures—in a much more sophisticated way than Dayan could do in the 1990s—how the diffusion of the different "small" digital media is part of the process of changing life in the diaspora without conceptualizing this process in a one-sided way as being the "effect" of these media. However, in this case she has a tendency to overlook the still important "big" media, instead seeing it as diasporic media, like Bailey, Georgiou and Harindranath (2007), or origin (culture). Indeed, "communicative connectivity" does not just occur as a specific moment in which digital media is appropriated, but is a more general process of media communication (Hepp et al. 2008).

From our perspective, it is thus essential to analyze the communicative connectivity of the migrants in its entirety if we want to make any statements on their (media) culture. In this sense, we want to extend the concept of the "connected migrant" to that of the "mediatized migrant". With this term we reflect the all-embracing "mediatization" (Krotz 2009; Lundby 2009) of diasporic lives. Like other people in Europe or North America, migrants live highly media-saturated lives. This means that we can only understand the everyday life of migrants if we focus on the interrelation between different media and everyday practices. This includes (traditional) mass media as well as (new) media of personal communication. In total these media constitute a "media environment" that also produces certain social possibilities and restrictions (Meyrowitz 1995, 51). From the perspective of an individual person, a media environment becomes concrete in the form of "media repertoires" (the totality of the interrelation of the media used) (Hasebrink and Popp 2006). Considering this from the point of view of the communicative connectivity of mediatized migrants, it is possible to see that the totality of interrelated mass-mediated and personal communicative connections are not articulated by the appropriation of a single media, but by the interference of different media in the whole repertoire of a person.

Referring back to our earlier point we can now capture the term "diasporic media culture" more clearly: fundamentally, diasporas are a specific form of cultural thickening. While being related with an (imagined) origin (culture), they are also marked by their own form of identity and community that arises from living abroad. For example, the self-image of a German Turk is different from the self-image of a Turkish person travelling as a tourist to Germany. Cultures of diaspora are *media* cultures inasmuch as their main resources and needs are communicated by the media. At this point, there is no difference if compared to other present media cultures. However, the specificity of diasporic media cultures can be seen in the fact that "territoriality"[1] is not constitutive for them. In contrast to national media cultures, which operate with the imagination of a territorial extension of national states, we must understand diasporas much more as networks of local groups being dispersed across different states but at the same time articulated by a shared meaning that can be articulated through the use of media. Because of this deterritoriality, media in total have a special status for migrants and to a certain extent it is only possible for *mediatized* migrants to articulate a shared cultural thickening of diasporas between an (imagined) origin and the migration context.

Nevertheless, we have to be careful with such an approach to diasporic media cultures, and reject any tendency to construct them in a homogeneous way. Aksoy and Robins (2000) discussed this in their research on Turkish migrants in London. Reflecting on this research, Robins argues that we should not idealize the "particular transnational media culture" (Robins 2003, 187) of migrants as diasporas because this concept is related to a problematic description of migrants as members of an "imagined community" (Anderson 1983) with a clearly defined cultural identity. In contrast to this, Robins argues for a "move from the level of the 'fictive unity' to that of the 'empirical people'" (2003, 198). The research undertaken by Robins and Aksoy gives an understanding of their "intellectual and imaginative mobility" (Robins 2003, 201). The media appropriation of migrants is much more differentiated when considered as the homogenizing concept of the diasporas rather than an imagined community of shared identity indicators.

One should not underestimate the importance of these empirically based arguments. As nations are not conflict-free homogeneous entities with the same identity for all of their members (Hall 1993), it would be misleading to conceptualize diasporas as such homogeneous phenomena. However, the critique falls short from the moment it switches to the singularity of the (subjective) experiences of "empirical people" (Robins 2003, 198). While it is highly necessary to consider that a community ("Vergemeinschaftung") is a group of social relationships based on the subjectively felt sense of community spirit that has to be reconstructed in this perspective (Weber 1972, 21), it is problematic to conclude from the *variety* of subjective experiences the non-existence of a diasporic community or identity. Therefore, the arguments put forward by Aksoy and Robins (2000) remind us to have the (conflictual) multiplicity of diasporic media cultures in mind, without neglecting cross-sectional analyses of these cultural thickenings and their specific patterns of thinking, practice and discourse. For us it is fruitful to conceptualize the diversity of diasporic media cultures not just as individuality and subjectivity. Rather it seems to be helpful to understand their differentiation along different "types" of cultural identity and communicative connectivity. With such an approach it is possible to reflect on another aspect of the present discussion about media and migration, which is the question of to what extent the communicative connectivity of the diaspora establishes an "everyday 'transnationalism'" (Robins 2003, 203) or "cosmopolitism" (Georgiou 2007, 16)—or to what extent this is a normative ornateness of the occasionally highly precarious experience of migration.

METHODOLOGICAL APPROACH: CULTURAL IDENTITY AND COMMUNICATIVE CONNECTIVITY

The study we present is based on multilevel research. While our original research question was focused on the specific potentials of digital media (mobile phone, e-mail, chat, WWW, etc.), for the communicative connectivity of the diaspora our field work quickly revealed that this potential can only be described in the whole context of media appropriation, including traditional mass media. The communicative connectivity of a person is not constituted by the appropriation of a *single* medium (for example television), but by the interrelation of various media in the whole repertoire of a person.

In addition to the secondary analysis of survey data on media use, our research is based on media ethnographic research using a transcultural comparative design. This aims to comprehensively reconstruct the cultural identity and communicative connectivity in the subjective perspective of the migrants in order to arrive at a grounded understanding of media appropriation in the diaspora. Our concept of media ethnography is not focused on the "thick description" (Geertz 1994) of migrants' life-worlds. Rather we understand media ethnography as an "ethnography of people who use, consume, distribute or produce media" (Bachmann and Wittel 2006, 183). Accordingly, media ethnography in general does not operate with long-duration stays in one field (Lotz 2000). Rather, "accumulated ethnographic miniatures" (Bachmann and Wittel 2006) (the combination of a number of different short stays, observations and interviews) or a "virtual ethnography" (Hine 2000) or "netnography" (Kozinets 2010) (ethnography approaches solely oriented on the Internet) are used to make propositions on specific ways of media appropriation.

The Moroccan, Russian and Turkish migrant communities were selected for this study first in relation to the region or origin (countries on different borders of the E.U.), second related to their size (with the Turkish and Russian diasporas two of the biggest, and with the Moroccan diaspora one of the smaller migrant communities) and third in relation to the temporal phases of migration (with a high migration phase of the Turkish diaspora between the seventies and nineties, a high migration phase of the Russian diaspora in the nineties and a more continuous mobility in the Moroccan diaspora).

All data was collected in the two German cities of Berlin and Bremen and their rural surroundings, and included more than thirty persons per diaspora. All respondents were selected by the principle of diversity reflecting age, duration of residence in Germany, education, gender and according to their self-definitions as members of one of the migrant groups.[2] Each person was interviewed about their migration experience, identity and media appropriation, asked to draw a network map and, finally, to keep—as far as possible—a two-week media diary. The network maps are free drawings of the person's communicative network, which was explored during the interview. These maps were used in triangulation with the interviews to capture the structures of communicative networks, while the media diaries made it possible to reconstruct the processes of communicative networking over two weeks. Additionally, we collected different forms of material documentation (besides other photographs of private and public locations of media use). The whole data was analyzed by a process of (open) coding oriented to the "grounded theory" (Glaser and Strauss 1999). This coding was guided by our research aim to describe everyday patterns of media appropriation in these three diasporas as well as their communicative connectivity in relation to dominating forms of cultural identity.

Where we use the term "pattern", we understand by this "cultural forms" in the general sense of cultural analysis (Hepp 2009b). The term "pattern" tries to express the idea that a cultural analysis should not analyze just the single "thinking", "discourse" or "doing" but should depict, based on an analysis of different single phenomena, the typical "way" of "thinking", "discourse" or "doing" in a certain cultural context. In other words, a cultural pattern is a specific form or type highlighted in cultural analysis. In this sense our research is oriented to building a qualitative typology "transculturally". This means we avoided a categorization that is, from the beginning, related to just one diasporic community, and instead tried to develop a categorization across them.

As a result of this coding process we obtained a system of eighty-six single codes. In relation to our research question, these codes can be structured in the two key categories "cultural identity" and "communicative connectivity". The category "cultural identity" captures four sub-categories (via which further sub-categories of identity and belonging are systematized): details about the person; self-image; migration experience; locations of living. "Communicative connectivity" is related to seven sub-categories (via which we arranged categories on a further sub-level): media offers; media arrangement; communication network; contents; locations of use; media experience; roles of intermediaries.

TYPES OF APPROPRIATING CULTURAL IDENTITY AND COMMUNICATIVE CONNECTIVITY: ORIGIN-, ETHNO- AND WORLD-ORIENTED MIGRANTS

Combining our two key categories—cultural identity and communicative connectivity—we can distinguish across the researched diasporas three types of appropriation: "origin-oriented", "ethno-oriented" and "world-oriented" migrants (see Figure 11.1). While these three types have a specific characteristic for each diaspora and are spread differently across them, they offer in total the possibility to describe the potential of media for migrant.

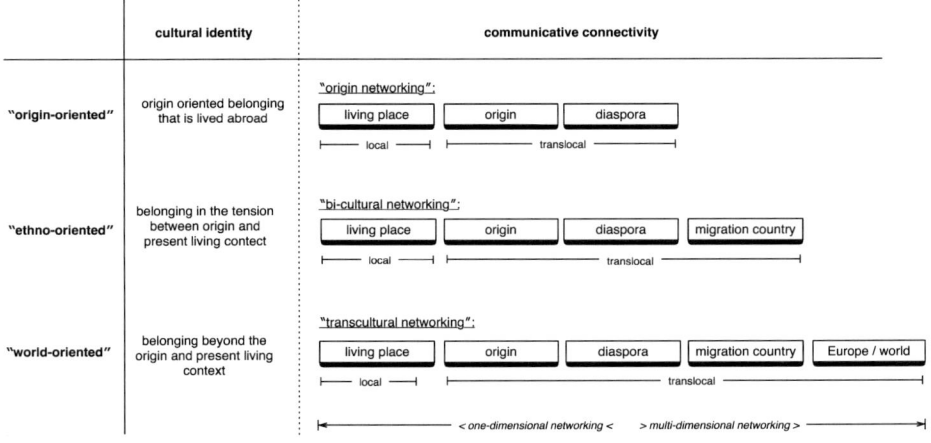

Figure 11.1 Types of appropriation.

The names of the three types are based on the subjective positioning of cultural identity, i.e. the self-definitions of the interviewed migrants. We found a correspondence between cultural identity and migrants' communicative connectivity indicated by our analyses of their network cards and the information obtained from the interviews. The figure above can only partly visualize this, as across the different types there is also a change in the quality of communicative connectivity. This means that the meaning of communicative connectivity to the present living place, the origin, diaspora, etc. is changing from type to type. The difference between local and translocal communicative connectivity is important for this. *Local* communicative connectivity means the connectivity at the present living place, i.e. the direct, everyday living environment. This connectivity is built up—*beside* face-to-face communication—via media of personal communication (mobile phone, e-mail, etc.) as well as media of mass communication (local newspapers, radio, etc.). Translocal communicative connectivity means the overall connectedness of the living place for which—besides travelling[3]—different media are of relevance. This translocal communicative connectivity can be oriented to the origin, the diaspora, the migration country or further entities (like Europe, for example).

Simplified, we can say that for *origin-oriented* migrants, their subjective sense of belonging marks their life "in exile". This subjective belonging can (but not must) be based on a socialization in the region of origin. Especially for younger migrants who grew up highly focused on their migrant community in Germany, the origin-belonging is based totally on imagination and some experiences from holiday travelling. The cultural identity of origin orientation is characterized in our interviews via formulations like, "I am Moroccan . . . It does not matter where I . . . live" (Fatih, male, 28 years, Moroccan diaspora);[4] via self-definitions like the relation to "the Soviet [sic!] culture" (Pawel, male, 59 years, Russian diaspora); or just a self-description as "Turks" (Feraye, female, 35 years, Turkish diaspora). It is not necessarily a problem for origin-oriented people to live abroad, but troubles begin especially when they have the subjective feeling that they are not accepted with their own identity or if they have to live in precarious economic situations.

This orientation of cultural identity goes hand in hand with a specific communicative connectivity that can be described as "origin networking". While there is an intensive local communicative connectivity, mostly in relation to other members of the same diaspora, we find additionally various translocal communicative relations to the region of origin. The media repertoire of this type is oriented to origin networking. For example, Noureddin (male, 27 years, Moroccan diaspora) uses the telephone to maintain contact with his relatives in the country of origin. Additionally, the communicative network of origin-oriented migrants is focused on migrants in their locality with the same origin. And for mass media, we can say that origin-oriented migrants have the tendency to use media offers of origin (satellite television, radio or newspapers)—and when they use German mass media, then quite often they use local or regional offers that ensure the communicative embedding in the present living context. On the whole, the media environment of origin-oriented migrants differs in relation to age, education and economic situation—however, across such differences it is marked by the use of origin networkin*Ethno-oriented* migrants differ. The name of this type indicates that their feeling of belonging is marked by a feeling of a dual identity, that of the country of origin and that of the present (national) migration context. Typically, these migrants call themselves "German Moroccans", "German Turks" or "German Russians". The center of belonging is much more the part of the diasporic community which is located in Germany than their country of origin. With "ethno-oriented" we want to express that migrants of this type perceive questions of ethnic positioning as the central aspect of their own identity. Mahmut (male, 30 years, Turkish diaspora) says he experiences "Turkey as [his] home as well as Germany." Amir (male, 57 years, Moroccan diaspora) characterizes himself as a "hybrid . . . of both cultures" and Valerij (male, 68 years, Russian diaspora) says, "I cannot count myself as part of this German culture alone." The communicative connectivity of the ethno-oriented migrants can be described as "bi-cultural networking". This term emphasizes that the communicative connectivity occurs in the tension between two (imagined) cultures. In the intersection of both, the diaspora is constituted as its own cultural thickening. Certain moments of opening happen especially locally at the living place: the local communicative networking includes, besides members of their own diaspora, other migrants or Germans. As an example we can quote Victoria (female, 47 years, Russian diaspora), who lists as part of her circle of acquaintances "not only persons from Russia" but also "very pretty girls who came from Turkey. Louisa from Brazil . . . Evan from Moldova . . . He has lived several years in Italy, because of that I don't know where he is exactly from."

Compared with origin-oriented migrants, the translocal communicative connectivity is also much more complex and embracing. There are communicative relations to the (imagined) origin. Additionally, there is an intense communicative networking in their own diaspora *with immigrants from other countries*. This communicative connectivity is carried out by a media repertoire that includes (digital) media of personal communication as well as mass media. Dominant forms are—besides local offers—diasporic, German and origin offers. In this sense, Aysen (female, 44 years, Turkish diaspora) says in relation to her reading repertoire, "We buy a Turkish and German newspaper each day to know what happens in Turkey and what happens here in Germany and here in Berlin." This corresponds again with the translocal personal communicative network of the ethno-oriented migrants, whose range includes German and their own origin, as well as their own family and diasporic friends.

· Yet another pattern of cultural identity and communicative connectivity is characteristic for the *world-oriented* migrants. The name of this type is borrowed from our interviews in which some of the respondents titled themselves "world humans" (Gökce, female, 33 years, Turkish diaspora) or as "Europeans" (Danil, male, 24 years, Russian diaspora). In so doing, they express a form of identity which is situated beyond the ethnic-national (in relation to the origin as well as in relation to the present migration country). Therefore, the term "world-oriented" expresses a subjectively felt belonging that is situated—on whichever level that might be—beyond the national. Pictures of the nation—the German or nation of origin as well as the bi-cultural tension between them—are encroached upon, and the supranational Europe or even the status of being human is the main point of reference.

This belonging goes along with a specific communicative connectivity that we can describe as "transcultural networking". This denomination clearly illustrates that we can compare the inclusion of origin and migration context within the communicative networking of ethno-oriented migrants. However, as opposed to this, the extent of communicative networking is much more comprehensive and has a tendency toward the European and the (imagined) global, or more concretely, to an extension of the communicative network across various countries and cultures. The media repertoire is manifold. Besides different mass media, media of personal communication are important—and among these, especially digital media. World-oriented migrants use e-mail, (mobile) phone, chat and—often with a high intensity—the social Web (such as Facebook) to stay in contact with people in their network. Within their personal communicative network, family and diaspora members are highly relevant. Moreover, the communicative network includes a number of other persons with whom the contact is established in part via work and education, in part for private reasons, and is more or less intensively maintained. Thereby the migration experience is absolutely a potential for contact deployment. To quote as an example for others:

> I have two cousins in Paris—oh, no -one in Paris, one in Saint-Etienne, in France—and one cousin in Moscow—and I have so many friends abroad because we studied together in Morocco and all of us went out to finish our postgraduate studies. So there are many in France mainly, and in Canada because in the Quebec part in the French speaking part of Canada. (Inaya, female, 29 years, Moroccan diaspora)

In all, we can conclude that for each of the three fundamental types a specific interrelation of cultural identity and communicative connectivity is characteristic—an interrelation which we cannot interpret as a one-sided causality. This means that origin-oriented, ethno-oriented and world-oriented cultural identity or subjective belonging do *not* have any particular communicative connectivity *as an outcome*. Furthermore, origin networking, bi-cultural networking or transcultural networking *do not have the effect* of creating a specific cultural identity. Rather, we have to understand the interrelation between cultural identity and communicative connectivity as a *co-articulation* in which both are mutually reinforced. The origin networking strengthens the articulation of an origin-oriented cultural identity, and therefore a corresponding orientation of the communicative relations. A bi-cultural networking strengthens the articulation of a doubled (or hybrid) cultural identity, and therefore, in turn, an orientation to communicative relations between origin and migration country. A transcultural networking strengthens the articulation of a European or global belonging, and hence a focus on far-reaching transnational communicative relations. In all of these processes of co-articulation, different media of personal communication and media of mass communication go hand in hand.

SPECIFIC CHARACTERISTICS ACROSS THE DIASPORAS

The quotes with which we have substantiated our analyses already indicate that the types examined above are not spread equally across the Moroccan, Russian and Turkish diasporas and that there is a need to look at certain cases in detail. In this sense, we understand our typology not as an undertaking to homogenize the researched migrant communities. Rather, they are—as already discussed—just as manifold as national communities. Consequently, we have to grasp the typology as a starting point for describing the specificity of media appropriation within these diasporas, as well as in single sub-groups within and across them, in a comparative manner. While it is impossible to do this for all researched diasporas in this chapter (see for further analyses Hepp 2009a; Hepp, Bozdag and Suna 2011), we want to use this section to shed light on some basic differences.

The one hundred cases we have researched (selected by sampling the specificity of the researched diasporas) enable us to make some statements, albeit not generalized, regarding the tendencies we have found in our data. Presenting our data in the form of tables and figures might be unusual for a mainly qualitative research. Nevertheless, we do this because it allows us to substantiate some core arguments in a very condensed way.

First of all, when we reflect upon the spreading of the different types across the researched diasporas (Table 11.1) it becomes apparent that the ethno-oriented migrants dominate, while the world-oriented migrants are rarer, especially in the Turkish diaspora.

Table 11.1   Types in Relation to Diasporas (absolute numbers)

	origin-oriented	ethno-oriented	world-oriented
Moroccan diaspora (n=32)	9	18	5
Russian diaspora (n=31)	9	17	5
Turkish diaspora (n=37)	12	23	2
Total	30	58	12

Also, there are differences in relation to the age of type cases and their spread across the subjects we investigated (Figure 11.2). Especially origin-oriented migrants inside the Turkish diaspora tend to be older than ethno-oriented migrants. Striking is also the age of these world-oriented migrants—they are by far the youngest.

If one analyzes the respondents in relation to their education,[5] we find that the formal education of migrants of all diasporas does not always correspond to their present occupation. It is a recurrent experience that their original formal qualifications are not acknowledged in Germany and that they have to work in non-adequate fields. Reflecting on this limitation, it is nevertheless striking that the criteria of formal qualifications alone does not allow an easy assignment to any one type. In the group of ethno-oriented migrants, all diasporas in fact include persons with very different formal qualifications. On the contrary, if one considers that most of the world-oriented migrants without school-leaving qualification still are in school, and that most of the world-oriented migrants without higher education qualifications are still at university, we can argue that world-oriented migrants are in tendency better educated than others.

Figure 11.4 demonstrates that the stereotype about the belief that origin-oriented migrants cannot speak German is not confirmed. Nevertheless, we can say that across all three diasporas the German language skills of the origin-oriented are the lowest. However, again we have to be careful; very often origin-oriented migrants have high language skills in their origin language, while ethno-oriented migrants—especially in relation to writing competence—have difficulties with their origin language, which partly limits their media appropriation (for example, in relation to chats in the origin language) and other possibilities. In addition, we consider that world-oriented migrants tend to speak and write a further language. This is

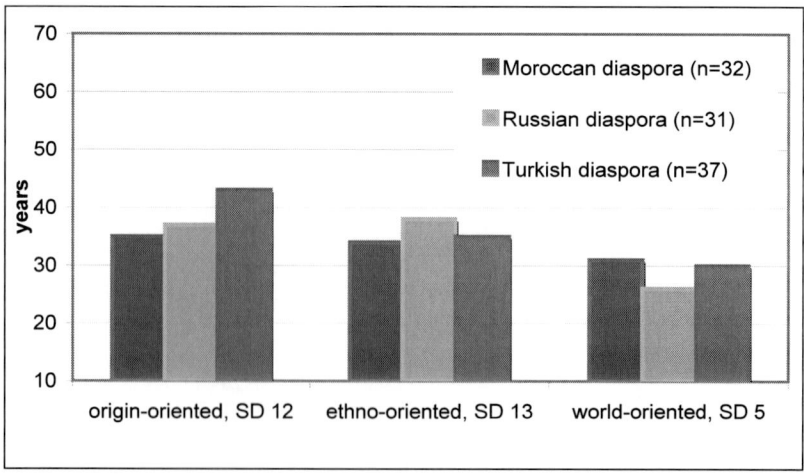

Figure 11.2   Age per type and diaspora (mean value).

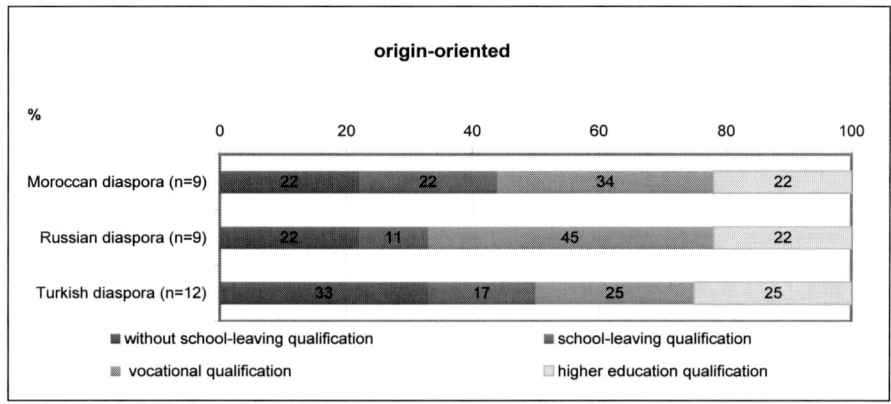

a.

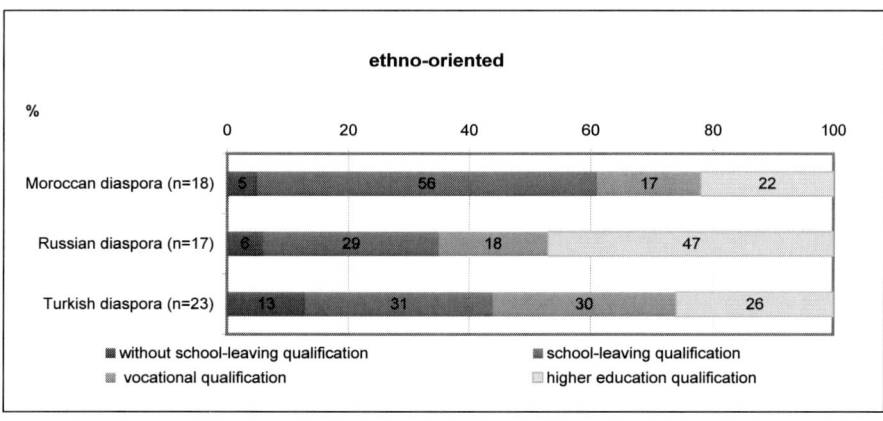

b.

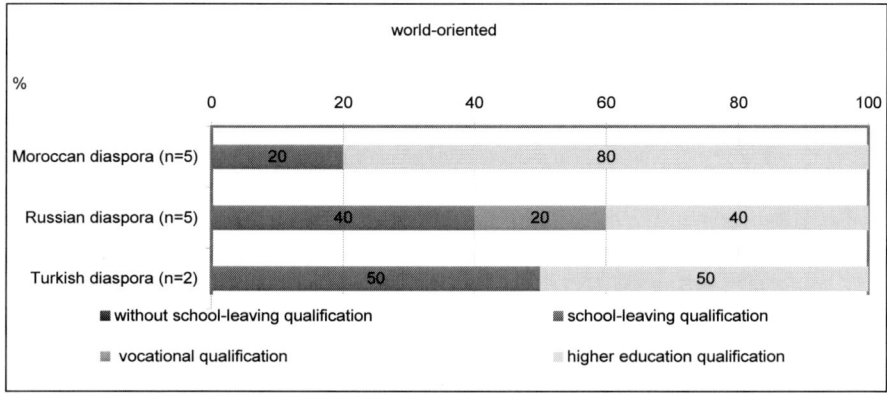

c.

Figures 11.3 Education per type and diaspora.

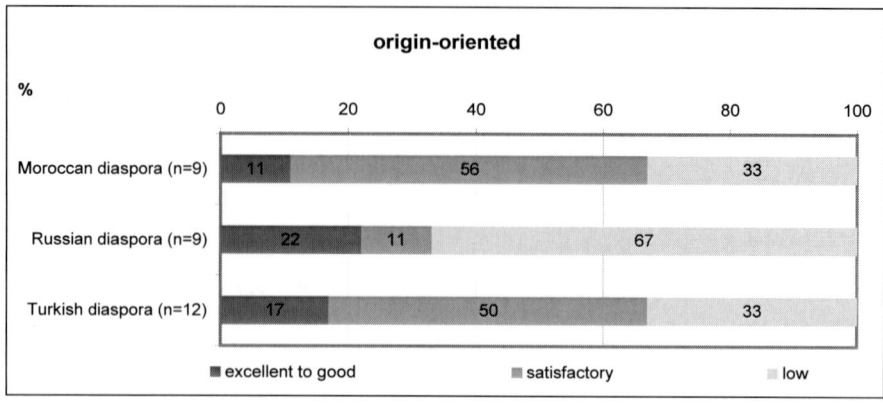

a.

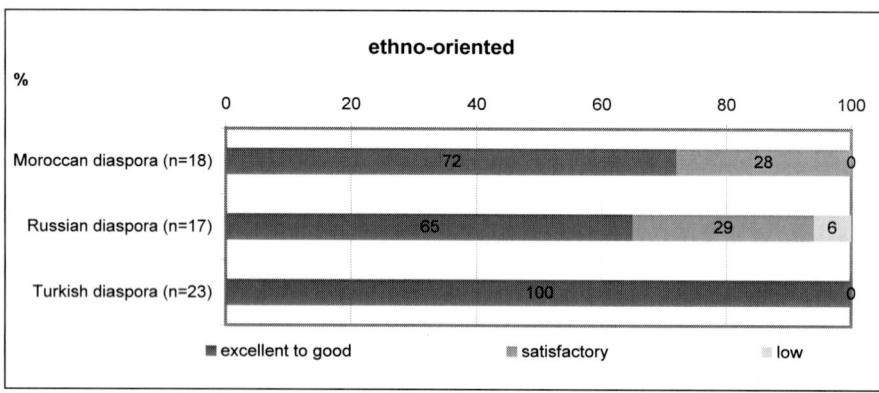

b.

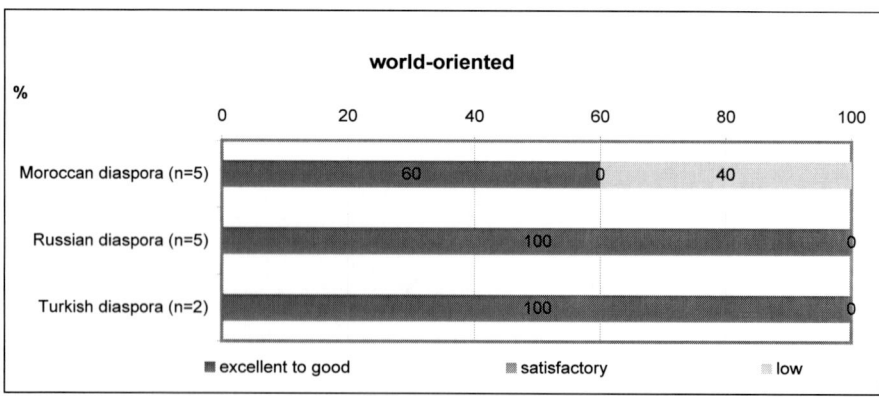

c.

Figures 11.4   German language skills per type and diaspora.

mostly English; in the Moroccan diaspora, also French, Spanish or—if the migrants are Berbers and Arabic is not a first language—Arabic.

These corresponding points become more striking if we focus more closely on our qualitative data, not just the interviews but also the network maps (see Figure 11.5). A first example is the origin-oriented Polina (Russian diaspora). Her network map clearly shows her origin and (local) diasporic connectivity via face-to-face and mediated communication. German is used primarily for institutionalized formal communicative contacts (for example, doctors or administrative bodies) or in a passive way for watching television. Other forms of communicative contacts are

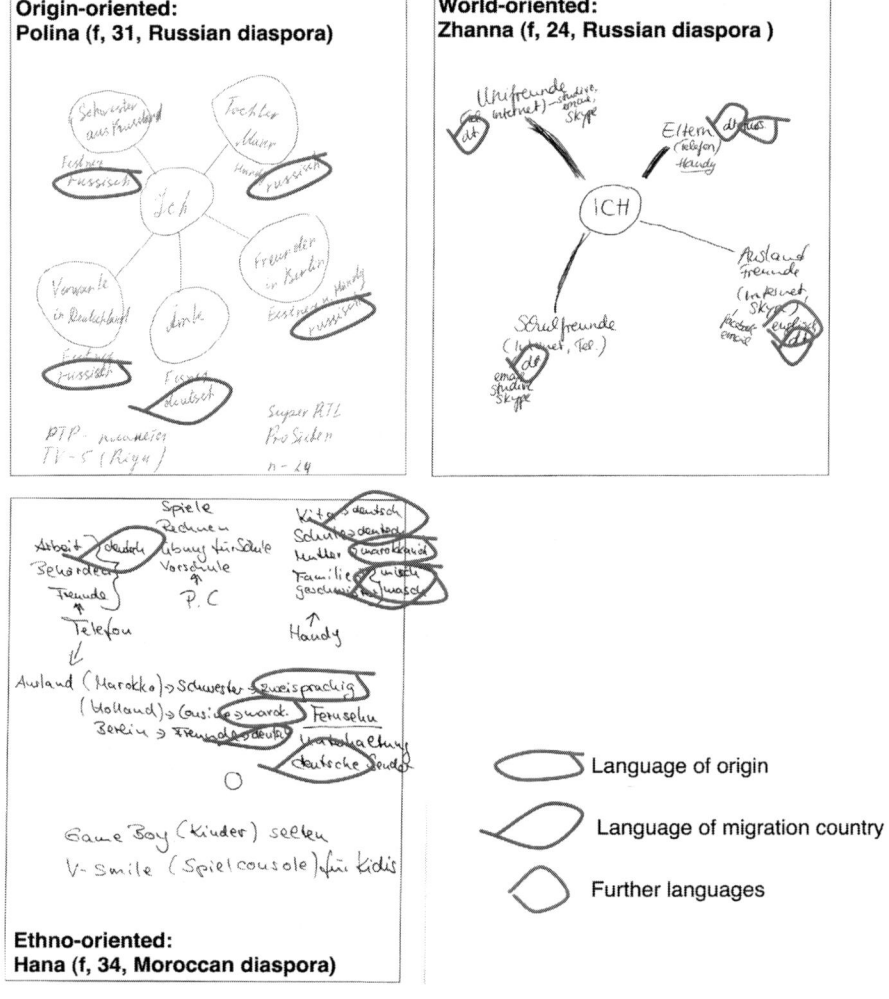

Figure 11.5 Communicative connectivity and language.

dominated by Russian. The bi-cultural networking of the ethno-oriented Hana (Moroccan diaspora) takes place in her origin language as well as in the language of her migration country, or—as Hana says in her own words—in a "mishmash" of both, especially within the communication in her family. The world-oriented Zhanna (Russian diaspora) is widely communicatively connected via different media. Reflecting on the issue of language, the network map shows the status of English for her communicative networking with different countries. Ethno-oriented, but especially world-oriented persons like Zhanna, have an understanding of the potential their multilingualism provides, such as for opening up certain job opportunities. But also the language competence of an origin-oriented migrant can have potential, for instance together with technical skills, for configuring computers in the language of origin, like in the case of Genadij (male, 30 years, Russian Diaspora), who earns his living this way.

This said, we can conclude that we find certain tendencies in relation to the types distinguished by us across the researched Moroccan, Russian and Turkish diaspora. In particular we found that world-oriented migrants are younger and better educated than others, although they seem to be more rare in the Turkish diaspora. However, there is no clear relation between any one diaspora and any one of these types.

DIASPORIC MEDIA CULTURES: RESEARCHING MEDIATIZED MIGRANTS

The analyses that we presented up to this point, we suggest, affirm an empirical substantiation of our theoretically developed argument that the multiplicity of diasporas should not lead us to dismiss a cross-sectional approach in analyzing the communicative connectivity of migrants. The achievement of this is not intended to replace further detailed analyses of diasporic media cultures or of migrants within these groups—in our case, of the Moroccan, Russian and Turkish diaspora. Rather, the empirical data we analyzed might be an important starting point that helps to capture three important aspects of this research area.

First, we can describe the conflicting variety of different identity orientation and related communicative connectivity within diasporas. It is clear that the origin-oriented and world-oriented migrants of the same migrant community might have completely different views of the world than, for example, ethno-oriented migrants (and also different media use).

Second, we assume such a multiplicity of views can be either individually subjective or typical for any particular diaspora. As a result it must be understood that there might be different possible orientations of mediatized migrants in the present.

And finally, third, we can empirically ground the discussion on the "cosmopolitan potential" of the diaspora, which is often outlined in an absolute way. While we find situations of "everyday cosmopolitism" in the cultural identity and communicative connectivity of the persons we have called world-oriented, this differs in respect to the types of the origin- and ethno-oriented migrants, who reject this kind of orientation. In this sense we understand our typology also as stimulation for a more differentiated discussion on the cosmopolitan and transcultural potential of diasporic media cultures.

It is recommended here that when researching "mediatized migrants" empirically, we must keep all three points in mind. While being a migrant in the present is a mediatized experience, a result of this is not a certain "media logic" on the status of being migrant. Rather we have to analyze very carefully in which way the molding forces of the media are relevant for migrant lives. Hopefully, our typology suggestion will stimulate such research.

NOTES

1. We use the term not for the general, physical dimension of localities, but in a more differentiated sense of describing the closed socio-geographical entity, as it is for example typical for nation-states.
2. In the case of the Russian diaspora this meant that persons from different countries of the former USSR were selected. Their self-definitions as part of the Russian diaspora were independent of whether the (present) country of origin is Latvia or Ukraine, for example.
3. For example, cheap flights for connecting diasporas.
4. All names are pseudonyms. The quotes are slightly familiarized to the British orthography and grammar to facilitate readability.
5. Figure 11.3 is conservative in the sense that only already-completed examinations/certificates are counted; that means, for example, that a person being in professional training is counted as a person with school-leaving qualifications but not with professional training qualifications.

REFERENCES

Aksoy, Asu, and Kevin Robins. 2000. Thinking across spaces: Transnational television from Turkey. *European Journal of Cultural Studies* 3(3):343–365.

Anderson, Benedict. 1983. *Imagined communities: Reflections on the origins and spread of nationalism.* New York: Verso.

Bachmann, Goetz, and Andreas Wittel. 2006. Medienethnografie. In *Qualitative Methoden der Medienforschung*, ed. Ruth Ayaß and Jörg Bergmann, 183–219. Reinbeck and Hamburg: Rowohlt.

Bailey, Olga G., Myria Georgiou, and Ramaswami Harindranath. 2007. Introduction: Exploration of diaspora in the context of media culture. In *Transnational lives and the media: Re-imagining diasporas*, ed. Olga G. Bailey, Myria Georgiou, and Ramaswami Harindranath, 1–8. New York: Palgrave Macmillan.

Cottle, Simon. 2000. Media research and ethnic minorities: Mapping the field. In *Ethnic minorities and the media*, ed. Simon Cottle, 1–30. Buckingham: Open University Press.

Dayan, Daniel. 1999. Media and diasporas. In *Television and common knowledge*, ed. Jostein Gripsrud, 18–33. London and New York: Routledge.

Diminescu, Dana. 2008. The connected migrant: An epistemological manifesto. *Social Science Information* 47(4):565–579.

Geertz, Clifford. 1994. Thick description: Toward an interpretive theory of culture. In *Readings in the philosophy of social science*, ed. Michael Martin and Lee C. McIntyre, 213–231. Cambridge, MA: MIT Press.

Georgiou, Myria. 2007. Transnational crossroads for media and diaspora: The challenges for research. In *Transnational lives and the media: Re-imagining diasporas*, ed. Olga G. Bailey, Myria Georgiou, and Ramaswami Harindranath, 11–32. New York: Palgrave.

Glaser, Barney G., and Anselm L. Strauss. 1999. *Discovery of grounded theory: Strategies for qualitative research*. New Brunswick, NJ: AldineTransaction.

Hall, Stuart. 1993. Culture, community, nation. *Cultural Studies* 7(3):249–363.

Hasebrink, Uwe, and Jutta Popp. 2006. Media repertoires as a result of selective media use: A conceptual approach to the analysis of patterns of exposure. *Communications* 31(2):369–387.

Hepp, Andreas. 2009a. Localities of diasporic communicative spaces: Material aspects of translocal mediated networking. *Communication Review* 12(4):327–348.

———. 2009b. Transculturality as a perspective: Researching media cultures comparatively. Qualitative Social Research Forum. http://nbn-resolving.de/urn:nbn:de:0114-fqs0901267 (accessed January 1, 2009).

Hepp, Andreas, Cigdem Bozdag, and Laura Suna. 2011. *Mediale Migranten: Medienwandel und die kommunikative Vernetzung der Diaspora*. Wiesbaden: VS.

Hepp, Andreas, Friedrich Krotz, Shaun Moores, and Carsten Winter. 2008. Connectivity, network and flow. In *Connectivity, networks and flows: Conceptualizing contemporary communications*, ed. Andreas Hepp, Friedrich Krotz, Shaun Moores, and Carsten Winter, 1–12. Cresskill, NJ: Hampton Press.

Hine, Christine. 2000. *Virtual ethnography*. London, Thousand Oaks, and New Delhi: Sage.

Kozinets, Robert. 2010. *Netnography: Doing ethnographic research online*. London: Sage.

Krotz, Friedrich. 2009. Mediatization: A concept with which to grasp media and societal change. In *Mediatization: Concept, changes, consequences*, ed. Knut Lundby, 19–38. New York: Peter Lang.

Lotz, Amanda D. 2000. Assessing Qualitative Television Audience Research: Incorporating Feminist and Anthropological Theoretical Innovation. In *Communication Theory* 10: 447–467.

Lundby, Knut. 2009. Media logic: Looking for social interaction. In *Mediatization: Concept, changes, consequences*, ed. Knut Lundby, 101–119. New York: Peter Lang.Meyrowitz, Joshua. 1995. Medium theory. In *Communication theory today*, ed. David J. Crowley and David Mitchell, 50–77. Cambridge: Polity Press.

Robins, Kevin. 2003. Beyond imagined community? Transnational media and Turkish migrants in Europe. In *Media in a globalized society*, ed. Stig Hjarvard, 187–205. Copenhagen: Museum Tusculanum Press.

Tomlinson, John. 1999. *Globalization and culture*. Cambridge and Oxford: Polity Press.

Weber, Max. 1972. *Wirtschaft und Gesellschaft: Grundriss der verstehenden soziologie*. Tübingen: Mohr Verlag.

12 ICT Adoption by Immigrants and Ethnic Minorities in Europe

Overview of Quantitative Evidence and Discussion of Drivers[1]

Stefano Kluzer and Cristiano Codagnone

INTRODUCTION

The use of information and communication technology (henceforth ICT) by immigrants and ethnic minorities (henceforth IEM) has become a topic of European information society policies following the e-inclusion declaration signed in Riga in June 2006 (Presidency of the E.U. Council of Ministers 2006). ICT refers here broadly to computers, Internet, mobile phones and other digital devices and their applications. With the aim of extending ICT use and its expected benefits to all segments of society, the Riga declaration set as one of its priorities to improve "the possibilities for economic and social participation and integration, creativity and entrepreneurship of immigrants and minorities by stimulating their participation in the information society" (Presidency 2006, 4).

In this work, IEM broadly refers to foreign citizens born outside of the European Union (E.U.),[2] also called "third-country nationals", and to their children, known as "second generations". Because the origin and rights of immigrants have been constantly evolving along with the formation of the E.U. and its changing borders, we shall clarify when needed the IEM groups we refer to.

RESEARCH AIMS AND QUESTIONS

In the light of the above policy developments and the limited evidence about IEM's use of ICT in Europe, this work has three aims tackled in the central sections of this chapter.

First, we review the (very few) quantitative surveys that have attempted to measure ICT adoption by IEM in some European countries. We question whether it is true, as often assumed, that the disadvantaged conditions suffered on average by IEM compared to the native population should result in a lower take-up and use of ICT. We also question whether the same and/

or different factors (age, gender, etc.) differentiate adoption levels among IEM people.

Second, we provide an analysis of ICT adoption and use by IEM in Europe, integrating the views of scholars who have studied the "connected migrant" with the results of qualitative research recently performed at the JRC IPTS mostly focused on France, Germany, Spain and the U.K. Here we question the drivers and enablers (and barriers) of ICT use among IEM and how they affect technology appropriation (see below).

Third, we illustrate the results of a survey of about 120 mostly publicly-funded initiatives promoting ICT use by and for IEM across Europe, also identified as part of the mentioned JRC IPTS study. We question in which areas such initiatives are contributing to ICT adoption and appropriation by IEM, and whether they reflect any clear policy strategy and the different immigrants' integration models followed in the past by national governments in Europe.

After briefly presenting the key concepts and theoretical framework behind our work in the next section, we address the above research questions and finish with some conclusions and recommendations.

THEORETICAL FRAMEWORK AND KEY CONCEPTS

The theoretical frame of reference inspiring our analysis has been fully developed and grounded in a growing, multidisciplinary body of literature elsewhere (Codagnone 2009). Here we illustrate its key elements, albeit very synthetically.

We go beyond the notion of digital divide as a matter of access to ICT and look at digital inequalities as resulting from different patterns of use and especially appropriation of ICT. The latter refers to the process by which individuals incorporate ICT in their daily practices of working, dealing with government, learning, staying in contact with friends, buying goods and services, getting information and participating in the public sphere. Appropriation conveys the idea that individuals do things through ICT that are meaningful. Whereas access is increasingly widespread, differentiation in uses and in the capacity to effectively appropriate ICT can reinforce or disrupt existing patterns of inequalities and social stratification.

This perspective is inscribed into a new body of theoretical and empirical literature that considers ICT as a new, determining dimension of social processes (DiMaggio and Bonikowsky 2008; DiMaggio et al. 2004). Such literature aims at developing a comprehensive theoretical framework to understand the role of ICT on life chances and social stratification patterns.

We argue that appropriation is more important than mere access because the possibilities offered by ICT are more similar to those of cultural goods such as information and education than to those of mass media and simpler technology. Historically, as access to education spread across vaster segments of the population, new forms of differentiation emerged. The same may occur with ICT, whereby some social groups may be better positioned to appropriate them and derive the desirable benefits. The skills and cognitive competence required to simply use a mobile phone, for instance, are lower than in the case of using the Internet to find the needed information and select among different sources (judging their credibility).

In light of the above discussion, we look at some quantitative data to broadly analyze the issue of access to ICT, but we also go beyond and tease out how they can favour social inclusion or exclusion processes among immigrants and ethnic minorities.

THE HIGH ICT TAKE-UP BY IMMIGRANTS AND ETHNIC MINORITIES

In Europe, no systematic cross-country quantitative survey on the take-up and use of ICT by IEM has yet been developed.[3] In recent years, a few ad hoc surveys on this topic have nevertheless been carried out in some countries, whose results we illustrate below.

In the U.K., three statistical surveys (DfES 2003; Ofcom 2007, 2008) have looked at the adoption of ICT among the main ethnic minority groups (EMGs, as they are called there) in the country. They show that take-up of digital media and particularly of mobile phones and the Internet is similar (earlier surveys) or higher (most recent survey) among those EMGs compared to the U.K. population as a whole.

Table 12.1   Digital Media Access by Ethnic Minority Groups in the U.K. in % (2007)

	All UK Adults	Indian	Pakistani	Black Caribbean	Black African
Digital TV ownership	82	83	89	81	82
Mobile phone take-up	85	90	91	88	95
Internet take-up (all)	62	75	72	64	69
Internet take-up (under 45 years old)	74	78	82	73	71
Willingness to get Internet (among adults who do not have internet at home)	15	25	35	30	30

Source: Adapted from (Ofcom 2008)

Table 12.1 provides the latest U.K. figures from a sample of twelve hundred EMG people.[4] It shows that Internet take-up will further increase among EMGs given their higher adoption propensity compared to the national average. When only younger people (under 45 years old) are considered, Internet take-up rates become higher and more similar across groups, showing that age and related group demographics[5] are key factors to explain the difference in ICT take-up levels across the different population segments (for instance, Pakistani in the U.K. are on average much younger than the other groups). Income and socio-economic position, household structure (presence/number of children) and skill level also contribute to variations in take-up levels of computer and Internet by different groups, but they have been found to be rather more important in shaping usage patterns (breadth of media use, length of time spend online, simultaneous consumption, etc.).

Beyond age and socio-economic factors, "to some extent ethnicity also emerged as a factor in its own right, for on average in some key aspects South Asian and black groups emerge as disadvantaged, particularly South Asian (Muslim) women" (DfES 2003, xvii). Lack of computer literacy combined with language and/or literacy difficulties are reported as important barriers to computer use for some groups.

A survey in Germany performed in early 2007 by ARD/ZDF Medienkommission about media adoption by three thousand people from six immigrant groups[6] provides partly similar results (Simon 2007). Take-up of mobile phones and computers was higher by IEM groups, while daily usage of the Internet was higher among the German population (except for Polish immigrants), including when the younger segments are considered. A significant age-related gap is visible from the last two rows both within native and immigrant groups.

Table 12.2   Availability and Use of Digital Media by Immigrant Groups and German Population in % (2006)

	German Population	All Migrants	Late Ethnic German Repatriates	Turkish	Polish
Availability of cell phone	86	91	89	93	91
Availability of computer/ laptop	69	76	79	76	78
Daily internet usage (all)	28	22	23	20	29
Daily internet usage (age 14–29)	46	38	n.a.	n.a.	n.a.

Source: Adapted from (Simon 2007)

Age and other factors affecting Internet usage within IEM groups show up again in Table 12.3, about Turkish immigrants. The number of regular Internet users decreases significantly with age; being born in Germany and mastering the language also seem to make a big difference, and a strong gender gap is clearly visible.

In Spain, the National Statistics Institute (INE) interviews for the annual "ICT in households" survey,[7] a statistically representative sample of the population. In 2009 it reached about 7% of "foreigners".[8] The share of non E.U. citizens among them is unknown. Nevertheless, results showed in Table 12.4 are consistent with those from the other studies discussed here and from a survey conducted in Catalonia (Spain) which identified more accurately the respondents' nationalities.[9] Thanks to higher growth rates, over a few years computer and Internet penetration levels among foreigners have reached, and sometimes overcome, those of national respondents.

Table 12.3   Internet Use by Turks in Germany (2006)

	All	Age			Gender		Place of birth		Knowledge of German language		
		14–29	30–49	> 50	male	female	Germany	abroad	good	medium	little
Basis	500	198	236	66	202	298	163	334	229	106	165
Regular internet users*	36%	62%	25%	9%	46%	25%	67%	23%	57%	36%	12%

* Rather than daily use as in Table 12.2, 'regular use' here refers to use at least once in the 3 months before the survey.
Source: Adapted from (Simon 2007)

Table 12.4   ICT Users in Spain in the Last 3 Months among Nationals and Foreigners in % (2004–2009)

	2004	2005	2006	2007	2008	2009
			PC users			
Spanish	49.3	52.1	54.2	57.3	60.9	63.5
Foreigners	40.5	51.0	51.3	55.4	61.9	59.7
			Internet users			
Spanish	40.6	44.3	47.9	52.0	56.7	60.0
Foreigners	34.5	46.5	46.7	52.1	56.8	58.0

Source: our elaboration on INE data

With respect to communication purposes, Table 12.5 (referring to Internet users only) shows that foreigners were more frequently users of Internet-based communication services, except for blogs.

Concerning other usages of the Internet, users in the two groups have similar patterns with respect to leisure and entertainment purposes, whereas for personal purposes[10] many more foreigners use the Internet to search for a job (31% in 2008, 44% in 2009) compared to Spaniards (respectively 21% and 26%). Higher Internet use for job searching among IEM shows up clearly also in the U.K. figures (Ofcom 2008). As seen in footnote **43** of Ofcom (2008), personal use of the Internet mainly refers to the respondent's social and economic participation in the (host) society. The above result might therefore be interpreted to reflect shortcomings in those processes for IEM people, e.g. limited access to banking and financial services which are pre-requisites for online banking and purchases.

Finally, we can refer to a report by the Netherlands Institute for Social Research (SCP) (van den Broek and Keuzenkamp 2008) which in 2004 interviewed about thirty-five hundred IEM from four groups (Turks, Moroccans, Surinamese and Antilleans) on several aspects of their daily life including media and Internet use. The results show differences in Internet use across IEM groups, at an overall lower level compared to Dutch nationals. They again show very clear gaps within each group related to personal factors such as age (broadly corresponding to differences between first and second generation immigrants), education and knowledge of the Dutch language. Commenting on this, the report's authors conclude that whereas ethnicity plays a role in other aspects such as social contacts and participation in sports, as regards media consumption differences across ethnic groups disappear or become very small after statistical control for compositional differences is applied.

Table 12.5   Internet Use in Spain for Communication Purposes in the Last 3 Months, % of Internet Users (2008)

	Spaniards	*Foreigners*
Telephone	8.3	25.2
Video/Webcam	16.7	42.6
Chat, online forum	24.7	38.8
Instant messaging	52.9	66.0
Read blogs	31.5	24.4
Manage own blog	9.4	9.3
Other	8.0	9.8

Source: our elaboration on INE data

A MOSTLY USER-DRIVEN PHENOMENON

Albeit with variations across countries and ethnic groups, the evidence presented above shows a high level of ICT adoption by IEM, despite their weaker average socio-economic conditions.[11] Although sampling and interviewing methods often take into account the diversity of the IEM population, the surveys considered above are likely to be partly biased in favor of IEM people who are already settled, who own a fixed line phone and who can speak the host country's language. Newly arrived immigrants and the most disadvantaged IEM people are likely underrepresented and would show lower results. The earlier quantitative survey in deprived areas in the U.K. (DfES 2003) and the case study of the low-income neighborhood of Neue Vahr Nord in Bremen, Germany (Hepp, Welling and Aksen 2009) confirm this view. Nevertheless, the available evidence shows quite strongly that the take-up, use and interest for ICT are high among IEM living in Europe.

Besides the younger age profile of the IEM population, a crucial factor seems to be the presence of ICT use motivations which are strong enough to overcome many adoption barriers such as cost, access and skills.

The need to communicate within dispersed social networks of family members, relatives and friends is likely the predominant motivation. The major change brought by new, almost ubiquitous and low-cost communications services (Internet and mobile telephony, text and instant messaging, etc.) is their incorporation in daily life, rather than their occasional use for special events.[12] Immigration is deeply transformed by this change according to some scholars, who even speak of the new figure of the "connected" (Diminescu 2005) or "interconnected" migrant (Ros 2008).

The high transnational and local mobility of newly arrived immigrants and of those involved in temporary or circular migration (Castles 2006) is enabled by the use of ICT (Vertovec 2007) and in turn makes it necessary to better cope with such a lifestyle. Many immigrant workers need to keep in constant contact with their employers, change frequently their residency and give their mobile phone number or e-mail as the only certain way to be contacted (Ros 2008).

The desire to have up-to-date information about sports, politics, social life and culture from the home country or to access online services specifically targeted to Diaspora communities is another powerful motivation for the use of online media by IEM (Benítez 2006; Bernal 2006). The desire to access online content and information services from the host country is also a driver of ICT use for many IEM people. IPTS studies found mixed evidence on the importance of home vs. host country online content: almost exclusive orientation towards own language/country news, music and online social networking among Tamils in Paris; a similar, but less extreme situation with Ecuadorian immigrants in the rural area of Vera, Spain; mixed orientation among Bulgarians in Spain, young people from Maghreb in France and Polish people in Germany (Diminescu et al. 2009; Hepp, Welling and Aksen 2009; Maya-Jariego et al. 2009). Systematic differences between first and second generations (more oriented respectively towards home and host country online content and media) have been reported by other authors (Panagakos and Horst 2006; Peeters and D'Haenens 2005).

This leads us to a last but no less important point: the Internet and related services are often needed and used to explore and "bridge" into an ever more digital and networked European society. As an example we can mention the use of computers and Internet at home for children's education; this has been found as an important and distinctive motivation for ICT adoption by IEM people compared to all adults in the U.K. (Ofcom 2007).

More controversial is the interest for and use by IEM of online public services in their host countries. Very limited existing evidence shows that e-government services in Europe seldom target IEM customers; these in turn are very reluctant to use online administrative services. According to a French case study, "When it comes to sensitive issues or complicated procedures: people need someone to talk to. On the Internet, the migrant is afraid of not understanding and she does not entirely trust the system: she fears that the request (documents, registration etc.) will not be taken into account. This is even more so the case when there is a language barrier" (Diminescu et al. 2009, 35).

Inclusive e-government initiatives aim precisely to break this type of vicious circle, which also affects many potential service users among native citizens. However, such initiatives are not yet so common, especially in a cultural diversity perspective.

MARKET DRIVERS AND PUBLICLY-SUPPORTED INITIATIVES

Facing the strong ICT use motivations among IEM illustrated in the previous section, private suppliers and service providers in the host country and international markets seem to have been in general able to meet most of the ensuing demands, especially for communication services and for basic ICT access and use. This statement stems from a deduction and an observation.

The high ICT take-up levels (and growth trends) among this user group can reasonably be ascribed to market-based interactions with suppliers, because as discussed below the number of public and non-profit initiatives supporting ICT adoption by IEM people is low.

The observation on the contribution of market forces is that along with mainstream ICT suppliers one finds many Internet/phone shops or cafés (their name changes in each country) almost always set up by IEM entrepreneurs themselves. These provide low-cost telephone and Internet access services often to the benefit of specific ethnic communities.[13] They compensate for the lack of home PC and Internet access suffered by many disadvantaged IEM people, but they also respond to the need for greater privacy, autonomy and

Table 12.6 Internet Use in Spain in the Last 3 months: From Where (% of Internet Users) (2009)

	Home	*Work*	*School*	*Friends*	*Public library*	*Cyber Centre*
Spanish	82.2	44.7	14.6	28.1	10.0	6.6
Foreigners	67.6	18.1	9.8	21.3	6.7	32.1

Source: our elaboration on INE data

peer socialization felt especially by younger people. Table 12.6 shows the importance of cyber centers (as they are called there) for the foreigners living in Spain.

Beyond profit-oriented suppliers, IPTS research (Codagnone, Kluzer and Haché 2009) also points to an important role played by informal, often voluntary ICT "champions" and skilled young people in motivating and supporting users within IEM communities and promoting innovation. IEM associations and increasingly individuals are also active in providing online content and services for their own communities.

These self-help, bottom-up solutions are in most cases spontaneous, but at times they become part of more structured and visible initiatives (see later) aiming to compensate for a number of market failures. Some short-comings pertain to basic access and usage limitations. For instance, lack of infrastructure (e.g. broadband networks or public Internet access points) in deprived areas with a high concentration of IEM has been reported.[14] Also, as in the population at large some IEM people lack digital literacy and access opportunities, depending as we have seen on socio-economic (income and education) and demographic factors such as age and gender. Other shortcomings prevent a richer appropriation of ICT. Online content and services adapted or produced ad hoc for IEM users seem to be relatively limited, especially from public providers. Related to this, proficiency in the host country's language shows up as a barrier for basic and more advanced use. The growth of user capabilities[15] has not yet been specifically investi-gated among IEM people, but as for all users it cannot be taken for granted. To overcome these shortcomings different types of initiatives focused on ICT and addressing or developed by IEM people and groups have been implemented in Europe in recent years. Their total number is unknown, but about 120 of them have been surveyed by IPTS (Kluzer, Haché and Codagnone 2008) with the following main findings.[16]

Only 40% of the initiatives exclusively target IEM; the others are either general e-inclusion measures also addressing IEM people or measures involving them de facto, e.g. by operating in areas with a high concentra-tion of IEM. About 80% of the initiatives are carried out (often jointly) by third sector organizations (charities, voluntary and local community groups, etc.) and public entities. Most initiatives are funded mainly by public resources, whereas the private (profit-oriented) sector's involvement is limited.[17] About 45% of the initiatives have a local or regional focus, another 40% have a national scope and the rest operate at pan-European or international level (usually funded by E.U. programs). With respect to their aims and content, a first and most frequent type of initiative is represented by those which aim to help IEM using ICT by providing basic access and digital literacy opportunities, or more advanced ICT skills for specific jobs. The second type includes initiatives where ICTs are used directly and indirectly for helping IEM people meet their life situation challenges: by developing or adapting online content and services for IEM needs and by supporting intermediaries in the third sector (voluntary groups, charities, etc.) and service delivery actors such as civil servants, social care workers, doctors, teachers, media professionals and others working with/for IEM. In this group, only a few e-government initiatives and measures promoting IEM economic integration (entrepreneurship, job search, etc.) were found. The last group comprises initiatives for the appropriation of ICT by IEM users beyond personal needs for purposes such as: enhancing the visibility and "voice" of specific IEM groups and their collective organizations; facilitating the exchange of information and dialogue with the host society; documenting the collective memory and identity and/or the problems and achievements of daily life in local communities.

Significantly, more than 50% of all surveyed initiatives witnessed a leadership role or the active involvement of IEM people in the implementation and delivery process. In fact, the development Web 2.0 services and applications (social networking services, video and picture sharing services, blogs and so on) clearly show up also in this field, with the boom of user-driven content creation initiatives (the survey only selected a small sample of them).

Despite the important role of the public sector in the promotion, funding and implementation of these initiatives, the IPTS study could not find any explicit e-inclusion policy in Europe (at the national level) specifically devoted to IEM.[18] The above initiatives therefore in most cases stem from bottom-up pressures which coalesce around (local) public funding opportunities (mostly short-term) and the efforts of third sector organizations. These conditions create their own problems: the vast majority of third sector organizations actually lack adequate digital tools and capabilities; projects often suffer from limited visibility, few occasions to exchange and network with similar actions and from lack of adequate scale and sustainability.[19]

CONCLUSIONS AND SUGGESTIONS FOR FURTHER RESEARCH

The most important conclusion from the evidence provided is that IEM are in general eager ICT adopters: most of them already use ICT and those who do not show clear intentions of future adoption.

Age, education level and gender also affect take-up and use of ICT among the IEM population. They thus determine digital divides within each IEM group and, depending on the relative size of such groups, the overall take-up differentiates between IEM people and the native population. The similar and sometimes higher take-up rates among IEM people than natives has much to do with the younger average age of the former. Host country language proficiency is an additional distinctive factor which importantly limits or enables ICT basic use and appropriation by IEM people. On the other hand, socioeconomic status seems to be less of a barrier for IEM than for the native population. Beyond these findings, the paucity of quantitative data constrains a deeper analysis. This remains an interesting and largely unexplored research path.

Strong motivations, primarily related to communication needs, appear as the main driver of ICT take-up and appropriation by IEM people. This is a clear conclusion from both quantitative and qualitative studies and can be related to the mobility and dispersed social networks of many migrants. However, the evolution of IEM's communication patterns (with whom, how frequently, about what, through which services) thanks to new technologies and services is still poorly known and deserves a more systematic research. Deeper investigation is also needed of the consequences of being constantly and deeply connected for the daily life in the host country, for the duration and orientation of migration projects and for integration prospects in the host society.

Other drivers emerging from qualitative and (less clearly) quantitative evidence are online information consumption and entertainment (especially of music and video). IEM users' orientation towards content and services on/from the homeland, the host country and other sources is still poorly understood, except for the existence of generational differences. Again, the consequences on daily life, migration projects and integration prospects of the multiplication of content sources and socialization opportunities (also across ethnic boundaries) brought by the new social media are still little-explored and deserve investigation.[20]

New media projects at the local level with a bottom-up approach emerged as a promising type of intervention to enable ICT appropriation and enhance employability of young (often second generation) IEM people, and at the same time to promote a richer socio-cultural interaction with the host society. As these initiatives become more frequent, it is feasible and useful to better assess their impact and find ways to optimize them.

The need to better "navigate" into the institutional and increasingly digital side of the host society came out from our research as a rather weak driver of ICT use. Exceptions to this are children's education and the search for employment, which motivate ICT adoption mostly on IEM families' and workers' own initiative. On the other hand, publicly-supported ICT-related initiatives in an integration perspective (access to government services, economic participation, etc.) are few and mainly address intermediaries and front-liners of public service provision rather than end-users. The determinants of this situation and the viability of alternative approaches are unclear and deserve further investigation.

Finally, what came out clearly from the IPTS study is that the traditional different migrants incorporation models[21] followed by European states in the past and in other policy domains do not show up in the field of e-inclusion. Whether those distinct models are fading away and merging into a new European-wide incorporation approach, which information society developments seem to be both revealing and enabling, is a fascinating research hypothesis that we believe is worth exploring further.

NOTES

1. This work stems from the research project "The potential of ICT for the promotion of cultural diversity in the EU: The case of economic and social participation and integration of immigrants and ethnic minorities" carried out by the Joint Research Center's Institute for Prospective Technological Studies (JRC IPTS) on behalf of Directorate General Information Society and Media of the European Commission. Related background materials and other publications can be found at http://is.jrc.ec.europa.eu/pages/EAP/eInclusion.html (accessed August 2, 2010).
2. According to Eurostat (Vasileva 2009), in January 2008 there were 30.8 million foreign citizens living in the E.U. (6.2% of the E.U. 27 total population of 500 million). About 63% of them (almost 20 million) were non-E.U. citizens, 31% of these from other European countries. Foreign-born people living in the E.U., including immigrants who acquired new citizenship, reached 50 million in 2006.
3. The annual households panel survey harmonized under Eurostat and carried out by statistical offices of all member states, which is the main E.U.-wide source of data on ICT use by individuals and households, introduced for the first time in 2010 the possibility (as an option) to request the nationality and the country of origin of the respondents. At the time of writing, however, the results of this change were still unknown.
4. A total of thee hundred face-to-face interviews were conducted for each group, mostly in England, sometimes using translated versions of the questionnaire into the languages spoken by those concerned. The interview sample reflects the age profile, geographical spread and other features according to the 2001 Census.
5. Under-45s account for 62% to 83% of all adults in each of these EMGs compared to 52% of the general U.K. adult population. By contrast, 19% of all U.K. adults are aged 65 and over, compared to only 2% to 5% in each EMG.
6. Besides the groups in Germany, immigrants from Italy, Greece and the Balkans were also covered by the survey, each one of them with approximately five hundred interviews.
7. The INE Web site gives public access to the "ICT in households" survey data, which can be queried for different years and variables. http://www.ine.es/jaxi/menu.do?type=pcaxis&path=%2Ft25%2Fp450&file=inebase&N=&L=0
8. In 2006, about 4 million foreigners lived in Spain: less than 1 million were from the E.U. and over 3 million were non-E.U. citizens (including at the time from Bulgaria and Romania). We do not know if the panel of the "ICT in households" survey reflected this nationality mix.

9. The survey performed in 2006 (Ros 2008) confirms the high ICT adoption rates by some IEM groups and also significant variations among them: people from "E.U. and rest of Europe" and those from Latin America used the Internet (respectively 78% and 77%) and e-mail (respectively 72% and 68%) more than Spanish respondents (57% and 43%). However, among those from the "rest of the world" (mostly immigrants from Asia and Africa) only 45% used the Internet and 30% used e-mail.

10. The survey envisaged the following personal Internet usage options: read e-mail, information search, travel information, download software, read news, job search, health information, online banking, e-commerce, information search on education/training opportunities, doing online courses, learning purposes.

11. This emerges clearly from the U.K. survey's results when sub-groups are analyzed by socio-economic level: "There is greater take-up among EMGs (compared to UK adults in general) among both ABC1s (higher status) and C2Des (lower status). In particular the socio-economic group differences observed in Internet take-up in the U.K. population are less pronounced in EMGs. This implies that socioeconomic group is less of a barrier to take-up among EMGs" (Ofcom 2008, 21).

12. Mazzuccato, Kabki and Smith (2006) showed how new ICTs allow for making decisions and arranging at the transnational level important life events such as weddings and funerals.

13. Maya-Jariego et al. (2009) provide a rich description of Internet/phone shops and their Ecuadorian users in the town of Vera, Spain.

14. See the case study on public digital access policies in France (Diminescu et al. 2009) and the case study on the Neue Vahr Nord neighborhood in Bremen, Germany (Hepp, Welling, and Aksen 2009).

15. These capabilities typically depend on opportunities for continuous learning through informal support from family members, friends, acquaintances and colleagues at work (Hargittai 2007).

16. The survey was performed in early 2008 in the 27 E.U. member states plus Norway and found initiatives in all of them except Cyprus, Estonia, Poland and Slovenia. The most important initiatives based on duration and/or notoriety were identified and characterized by objectives and content.

17. An exception is represented by the ICT training and community development projects funded by Microsoft's Unlimited Potential Initiative.

18. A partial exception is represented by the U.K. government's work started in 2000–2004 by the Digital Inclusion Team of the Social Exclusion Task Force.

19. These problems emerged also from an overview of another 150 e-inclusion initiatives not necessarily related to IEM (European Commission 2007) and from a comparative study of public Internet centers in four countries (Rissola 2007).

20. An exploratory study of these aspects can be found in (Diminescu, Jacomy and Renault 2010).

21. We refer here to assimilationist, integrationist and multiculturalist models, which are presented and discussed in the annex of the paper written by Codagnone and Kluzer (2010).

REFERENCES

Benítez, José Luis. 2006. Transnational dimensions of the digital divide among Salvadoran immigrants in the Washington DC metropolitan area. *Global Networks* 6(2):181–199.

Bernal, Victoria. 2006. Diaspora, cyberspace and political imagination: The Eritrean Diaspora online. *Global Networks* 6(2):161–179.

Castles, Stephen. 2006. Back to the future? Can Europe meet its labour needs through temporary migration? Working Paper 1, International Migration Institute, Oxford.

Codagnone, Cristiano. 2009. *Inclusive innovation for growth and cohesion: Modelling and demonstrating the impact of eInclusion*. Brussels: European Commission.

Codagnone, Cristiano, and Stefano Kluzer. 2010. ICT for integration of migrants in the European society and economy. Seville: JRC Institute for Prospective Technological Studies.

Codagnone, Cristiano, Stefano Kluzer, and Alexandra Haché, eds. 2009. *ICT supply and demand in immigrant and ethnic minority communities in France, Germany, Spain and the United Kingdom*. JRC Technical Note No. 52233. Seville: JRC Institute for Prospective Technological Studies.

DfES (Department for Education and Skills). 2003. *The use of and attitudes towards information and communication technologies (ICT) by people from black and minority ethnic groups living in deprived areas*, Nottingham, England: Department for Education and Skills.

DiMaggio, Paul, and Bart Bonikowsky. 2008. Make money surfing the Web? The impact of Internet use on the earnings of U.S. workers. *American Sociological Review* 73(2):227–250.

DiMaggio, Paul, Eszter Hargittai, Coral Celeste, and Steven Shafer. 2004. Digital inequality: From unequal access to differentiated use. In *Social inequality*, ed. Kathryn Neckerman, 355–400. New York: Russell Sage Foundation.

Diminescu, Dana. 2005. Le migrant connecté: Pour un manifeste épistémologique. *Migrations Société* 17(102):275–293.

Diminescu, Dana, Mathieu Jacomy, and Matthieu Renault. 2010. Study on social computing and immigrants and ethnic minorities: Usage trends and implications. Seville: JRC Institute for Prospective Technological Studies.

Diminescu, Dana, Matthieu Renault, Sylvie Gangloff, Marie Amélie Picard, and Christophe d'Iribarne. 2009. ICT for integration, social inclusion and economic participation of immigrants and ethnic minorities: Case studies from France. Seville: JRC Institute for Prospective Technological Studies.

European Commission. 2007. European e-inclusion initiative: First contributions to the campaign "e-Inclusion: Be part of it!" Brussels: Directorate General Information Society and Media, Unit H3.

Hargittai, Eszter. 2007. A framework for studying differences in people's digital media uses. In *Cyberworld unlimited*, ed. Nadia Kutscher and Hans-Uwe Otto, 121–137. Wiesbaden: VS Verlag für Sozialwissenschaften/GWV Fachverlage GmbH.

Hepp, Andreas, Stefan Welling, and Bora Aksen. 2009. ICT for integration, social inclusion and economic participation of immigrants and ethnic minorities: Case studies from Germany. Seville: JRC Institute for Prospective Technological Studies.

Kluzer, Stefano, Alexandra Haché, and Cristiano Codagnone. 2008. Overview of digital support initiatives for/by immigrants and ethnic minorities in the EU27. In *JRC scientific and technical reports*. Seville: JRC Institute for Prospective Technological Studies.

Maya-Jariego, Isidro, Pilar Cruz, José Luis Molina, Beatriz Patraca, and Alain Tschudin. 2009. ICT for integration, social inclusion and economic participation of immigrants and ethnic minorities: Case studies from Spain. Seville: JRC Institute for Prospective Technological Studies.

Mazzuccato, Valentina, Mirjam Kabki, and Lothar Smith. 2006. Transnational migration and the economy of funerals: Changing practices in Ghana. *Development and Change* 37(5):1047–1072.

Ofcom (Office of Communications). 2007. *Communications market special report: Ethnic minority groups and communications services*. London: Office of Communications.

———. 2008. *Media literacy audit: Report on UK adults from ethnic minority groups*. London: Office of Communications.

Panagakos, Anastasia, and Heather Horst. 2006. Return to Cyberia: Technology and the social worlds of transnational migrants. *Global Networks* 6(2):109–124.

Peeters, Allerd, and Leen D'Haenens. 2005. Bridging or bonding? Relationships between integration and media use among ethnic minorities in the Netherlands. *Communications: The European Journal of Communication Research* 30(2):201–231.

Presidency of the E.U. Council of Ministers. 2006. Riga ministerial declaration on an inclusive information society, June 2006.

Rissola, Gabriel, ed. 2007. *Suturing the digital gash*. Barcelona: Editorial Hacer.

Ros, Adela. 2008. Interconnected immigrants in the information society. In *Digital diasporas*, ed. Andoni Alonso and Pedro Oiarzabal. Reno: University of Nevada Press.

Simon, Erk. 2007. Migranten und medien 2007. *Media Perspektiven* 9:426–435.

van den Broek, Andries, and Saskia Keuzenkamp, eds. 2008. *Het dagelijks leven van allochtone stedelingen*. The Hague: Netherlands Institute for Social Research (SCP), Sociaal en Cultureel Planbureau.

Vasileva, Katya. 2009. Citizens of European countries account for the majority of the foreign population in EU-27 in 2008. Luxembourg: Eurostat—European Commission.

Vertovec, Steven. 2007. Circular migration: The way forward in global policy? In *IMI working papers*. Oxford: International Migration Institute.

Theme 5

A Case Study

China, Its Internal Migrations,
Diasporas and Expatriates

13 Migrant Workers, New Media Technologies and Decontextualization

A Preliminary Observation in Southern China

Pui-lam Law

INTRODUCTION

Since the early 1980s, when China implemented economic reforms and opened its doors to the outside world, internal migration from rural to urban areas and from the western and central regions of the country to the eastern coastal region has been continuous. The floating population, predominantly migrant workers, has increased sharply from 30 million in the early 1980s to 225.42 million by 2008. This scale of internal migration is unprecedented.[1]

In the early 1990s, there was a sharp increase in the numbers of migrant workers of the hinterland going to the coastal regions such as Guangdong in the hunt for work. Most of these workers led a frugal life so as to save a large part of their wages to use for the welfare of their family, such as building new houses and paying for weddings. These migrant workers regarded working in the city as just a transient period in their lives, and their goal was to return to their home village. These returnees not only brought back economic resources but also alternate values, life goals, and a new perspective on life (Murphy 2002) which they learned and experienced in the cities. Our study[2] reveals that the development of new media technologies and the recent rapid diffusion of these technologies have also made the less developed regions more open to developments in the cities of the coastal regions. In other words, new aspects of city life and new values and ideologies have penetrated, via the new media technologies, the less developed villages. Nowadays, young migrant workers attracted by the city life have flocked to Guangdong, particularly to the Pearl River Delta, not only to hunt for jobs to improve their standard of living, but also to experience the life of the city. However, life in the factories has not been easy, and the exploitation of their labor, the implacable opposition between the migrants and the local workers and feelings of desolation and insecurity have caused some migrant workers to return to their hometown. Those migrants who are determined to stay often find that the production process in the factory has turned them into an alienated object, with satisfaction experienced only through the consumption process in their leisure time (Pun 2003). The workers have become new desiring subjects.

Among the leisure time consumption practices of migrant workers most noteworthy are the purchase of new clothes or cosmetics, TV sets, MP3 players, the latest model of mobile phone (including subscribing to mobile phone services), purchasing Internet service in Internet cafés or through the mobile phone and going to karaoke bars. Some migrant workers follow TV serials either in their rented apartment or in the common room in the dormitory, and some even go to Internet cafés to watch movies or TV serials. The mobile phone is used to send SMS (short messages) to relatives or friends even from the assembly line, and to log on to QQ (a kind of instant messaging which is very popular in China, and particularly among the migrant workers) via the cyber network. Internet cafés provide Web cams, allowing migrant workers to chat with their cyber friends. The purchase of the new media technologies and their services has become an important part of their leisure time consumption process. These technologies are digitally connected to one another. Popular music or songs from the TV serials or radio can be downloaded from the Internet or a fellow workmate's machine to an MP3 or mobile phone; these songs can then be sung in karaoke bars. In view of this, we consider that these emerging digitally all-encompassing media technologies are moving the workers into the cyber world insidiously, resulting in their decontextualization from the face-to-face world, leading to social disorientation or deviation from the social norms of the real world. We have studied the social consequences of the uses of the mobile and the Internet since 2003 and in this research, understanding generally focuses on the positive effects of these two new media technologies (Law and Peng 2006, 2007, 2008). However, recent developments regarding mobile phone and Internet practices have attracted our attention and our understanding has started to skew more towards the negative effect of the consumption of these digital media technologies. As we have only been aware of this phenomenon since 2008, this chapter presents our preliminary assessment on the negative influence of the uses of the mobile phone and the Internet.

The data referred to in this chapter was gathered in interviews, surveys and observations of workers, Internet cafés and street scenes conducted by us since 2003 in the Pearl River Delta. We have interviewed 89 informants, 45 male and 44 female. Among them, 79 are workers, 9 are factory proprietors and managers and 1 is an Internet café manager. The informants come from three townships in Dongguan city and one in Shenzhen. Some were introduced to us by the factory management, local villagers or by the workers themselves; others we met in the Internet café. The structured questionnaire survey was conducted in April 2006 in three industrial townships in Dongguan City in the Pearl River Delta: Tangxia, Taiping and Dongguan. Successful questionnaire interviews were conducted among

655 migrant workers. The 655 respondents were drawn from nine factories in three industrial sectors: metal manufacturing (four factories with 424 respondents, or 64.73% of the total interview sample), plastic products (two factories with 126 respondents, or 19.24% of the sample) and textiles and garments (three factories with 105 respondents, or 16.03% of the total sample). Using the data from the interviews, surveys and observations, the first part of the chapter will present the uses of the mobile and the second the uses of the Internet.

THE MOBILE PHONE

The number of mobile users had reached 659.8 million (40.8% of the whole population) by February 2009. The penetration rate is highest in the coastal cities. In Guangdong, where the Pearl River Delta lies, the number of mobile users reached 82.25 million. In 2003, the penetration rate of the mobile phone among workers in the Delta was around 50% (Law and Peng 2007); in 2006, the penetration rate had gone up to 65% (Ngan and Ma 2008). When we were doing our interviews in 2008 and 2009 in the Delta, we noticed that nearly all the workers had a mobile phone and that the mobile phone was viewed by the workers as a necessity. In recent years, the functions of the mobile phone have expanded; it is no longer simply a tool for making and taking phone calls. It can be an MP3 or MP4 player, a radio, TV, digital camera or PDA; it can also be used to surf the Internet or connect to QQ. Perhaps this may partially explain why the mobile phone is ubiquitous among the workers.

The mobile phone of course still serves its telephone function and is gradually replacing the landline phone. In 2003, workers were very cautious when using their mobile to contact their families. Before answering a call, they made a point of first finding out who the caller was. If the call was from their family, by prior arrangement, they would not answer the call on their mobile except in emergencies. Rather, they went to a public phone booth to call their families. This prudent practice was mainly because of the differences in cost. At that time, the rate for making a local call from a fixed phone was 0.2 *yuan* (US$1 was approximately 8.28 *yuan*), but it was 0.6 *yuan* per minute from a mobile phone; the rate for making a long-distance call from a fixed phone was also 0.2 *yuan* per minute, but around 1.30 *yuan* per minute from a mobile phone. In addition to the fact that mobile phone rates were higher than those for fixed phones, mobile phone network providers charged both the calling and answering parties, while fixed phone companies only charged the calling party. Nowadays, the price has dropped significantly and some service providers do not even charge for answering a call; making a national distance call costs around 0.4 *yuan* (US$1 is currently approximately 6.84 *yuan*), but discount phone cards bring this cost down even further.

Consequently, the huge crowds that used to gather around the public phones in the factories or on the streets have now disappeared. This has been evident in the villages where we have been conducting research for over ten years. We have not seen any workers using the more than fifty public phones installed on the streets, and we have seen only a few workers subscribing to a landline phone service in the telephone supermarket.

The mobile phone provides convenience, yet this convenience does not seem to increase the frequency with which workers get in contact with their families in their hometown. In the early 1990s, when landline phones were not easily accessible, workers wrote letters to their family, often weekly or even once every couple of days. In 2003, when landlines had become more popular, our interviewees often complained that there were not enough public phones for them to use (Law and Peng 2007). Thus one of the interviewees said, "There are not so many fixed phones here, so it is difficult to contact my family. I could write to them every day and still not know when something is happening there" (Law and Peng 2007, 130). Because landlines overcome the spatial constraint, they triggered a demand for instant news from the other end far away. When mobile phones became more widespread in recent years, the workers were freed from the spatial and temporal constraints when connecting to their family in their hometown. This means they can now call their family wherever and whenever they like.

Both our survey and in-depth interviews have revealed interesting usage data, especially with regard to the difference between actual usage and those who the workers thought were their first (most important) contact. In the survey, carried out in 2006, we found that almost 70% of the young workers aged below 30 would put their family members, such as parents, spouse or sibling, as their first contact. Being placed as first contact, however, suggests only that they are important to the interviewee but it does not mean that they are the most frequently contacted persons. Our in-depth interviews with the workers revealed that calls to the family were, indeed, infrequent with some calling once a week, some even once a month. The convenience brought about by the mobile phone had only led to an increase in contact with friends in the host cities rather than in calls to family; the workers call their friends more frequently than their family members in their place of origin. Furthermore we found that they actually called their family less than they did before, let alone write to them.

In the past, when landline phones were not popular, migrant workers wrote letters to their family. Recipients felt close to absent family members just from looking at the handwriting. When the telephone, particularly the mobile, became popular, workers ceased to write letters as the phone became more convenient and the price became more affordable. As far as mobile usage is concerned, a phone call helps bridge the distance between the worker and their family, whereas a short instant message (SMS) or an e-mail can seem detached and abstract. As Heidegger maintains (1992), through one's handwriting we can know more of the one who writes, such as one's character, one's state of mind; but typewriting provides only a signless cloud, it conceals the essence of the writing. Thus, the typed words on the screen of a mobile are mechanically represented and consequently conceal the character of the writer. Simply put, the advent of new media technologies may have the effect of further distancing the affection inherent in face-to-face interaction among migrant workers and their family members. If it is true, the family ties may then be weakening and this would in turn lead to the weakening of family values.

Migrant workers commonly use the MP3 and MP4 function of the mobile phone to unwind after a hard day's work in the factory. Nowadays, it is very convenient and cheap to get songs or music downloaded to the mobile phone. One *yuan* will buy ten songs or pieces of music from the download shops in their industrial zone. As it is cheap and easy to access, the workers seldom choose the songs or music themselves, despite the fact that they have listened to some beautiful pieces on the radio or TV. Mostly, they buy the top ten recommended by *Baidu* (the most-used Web searching engine in China). A worker talked about downloading songs to her mobile: "I seldom choose songs myself, the staff in the Internet café download them for me, and they always fill up my phone." When they do not like the downloaded songs or music, they just delete them and download new pieces. Their favorite songs or music are therefore saved in their mobile phone for a week or less. The nature of music appreciation among workers has changed considerably now that it is stored as a form of digital data and is easily accessible.

In the past, particularly during the early 1990s, music appreciation was a component of the workers' leisure activities. But of course, the workers could only enjoy the music or songs through the radio or music cassette players. The number of pieces they could access would therefore be far fewer than today. When we interviewed the workers at that time, we found that they could recite the lyrics and explain the meaning of their favorite songs and why they loved them so much. *Dagong* songs (worker songs) were particularly popular among the workers because they reflected their experience in the host cities. Some songs told of the happiness of the future, some about the hardship and suffering found in the workplace and some about the exploitation the workers experienced. Today, the workers can barely remember the lyrics of the songs they download to their machine even though they can access them easily. Favorite songs are mainly romantic ballads; *dagong* songs are no longer popular. This may reflect the identity of this generation of worker, but perhaps music appreciation has now become merely a kind of information input, rather than its former function of identity reflection. The short life span of the music on their mobile phones may mean that music and songs have become a type of data and are data only to the workers. Above all, these data are ephemeral. Music appreciation seems to be significantly dominated by the new media technologies. In light of the above discussion, it would seem that the consumption of new media can exacerbate workers' loss of orientation in the host city.

Another function of the mobile phone is that of providing workers with abundant job information (Law and Peng 2008). The following is a story of how a migrant worker found a job through the use of a mobile phone in early 2000:

> Once my friend called me from somewhere and asked me whether I could find a job for his close friend from Shanghai. And at that time, our factory was recruiting workers, so I informed him by telephone. As we did not know each other, when he came here we had to fix up a place to meet up through our mobile phones. If we hadn't had mobile phones, this kind of thing would not have happened. (Law and Peng 2007, 136)

With the advancement of the technologies and the improvement of service, the workers now can simply make calls and send SMS or QQ messages to their kinsmen or former workmates to find out about job opportunities in the Delta. The mobile phone has partially contributed to the increase of job mobility rates (Ngan and Ma 2008). With more access to job vacancy information, workers can choose a better job for themselves. As workers circulate job market information through their mobile phones, they feel freer to move from job to job, particularly during the years when there were a great number of factory vacancies in the Pearl River Delta. It seems that the dormitory regime (Pun 2005) and capital oppression (Chan 2001) have become history in the Delta. This freedom to move from job to job, however, has not necessarily improved the workers' lives significantly. We noticed that although mobiles could well empower the worker against capital's control, some workers changed their job rashly. In our interviews, some workers told us that they wanted to change their job because the food in the factory canteen was not good; some said that they had to queue up for lunch or dinner in the canteen; some complained that they could not tolerate spiders' webs in their dormitory; and some simply followed relatives and friends to other factories. Most of the workers we interviewed in recent years had changed jobs frequently. One worker told us that he had had eight jobs in two years. Only a few significantly improved their lot through changing jobs. Apparently, the information that was circulating on the cyber network contributed to a kind of horizontal mobility. But with the meltdown of the world economy in 2008, factories in the Delta lost business and a huge number of workers lost their job. This phenomenon clearly shows that the autonomy the workers felt that they had with regard to job improvement and choice of jobs is in fact ephemeral, if not a mirage.

THE INTERNET AND THE INTERNET CAFÉ

Internet penetration among the workers has increased significantly. But a computer is still an expensive item and not suited to dormitory life; in addition, some workers are highly mobile and obtaining an Internet account is therefore difficult. Thus, migrant workers usually go to an Internet café to connect to the Internet. According to our survey, over 36% of workers go to an Internet café five days a week and 21% go every day. In 2005, there were only 314 legal Internet cafés in Shenzhen but by 2006, the number had risen to over 1,000 (Law et al. 2007). The wide availability of Internet cafés makes Internet use very convenient for migrant workers. By the end of 2006, the number of Internet subscribers in Shenzhen was reported to be over 2.15 million (Law et al. 2007).

Two types of workers frequent the Internet café; one goes there for work and the other just for fun and excitement. Of the first group of workers, most are clerical staff, supervisors or foremen, and they go there to search for data and information for work and personal development. One foreman working in an electronic device factory said:

> I feel that the Internet is very useful. If I want to get data related to my work, I can use the Internet to get it. Without the Internet, it would be very hard to get this data . . . And I often use the Internet to get job information as well. (Interview with a male foreman in an Internet café in an industrial district of Shenzhen, 2008)

These workers are not frequent visitors of the café and they spend less than 50 *yuan* a month, some even less than 30 *yuan*.

For the other group of workers, the Internet is a means of entertainment. Some workers said that the Internet use made their lives more interesting:

> "The main function of the Internet is relaxation and entertainment."

> "If I feel bored, I go to the Internet café."

These workers go to the Internet café to watch TV serials or movies, listen to music or songs, update their blog, play Internet games or chat on QQ, sometimes with the Web cam on. They often do all these activities simultaneously. They usually spend 100 yuan to 200 yuan a month. When the workers go to the Internet café for whatever kinds of Internet use, it is a place for a juxtaposition of the real and virtual for them. It is a place where they can meet workers from different levels of the hierarchy of their workplace, from different factories and from different regions of the country, both in real and in virtual worlds. This juxtaposition can be satisfying; in their online interaction they are on an equal footing with their seniors and supervisors. This is shown in one interview in which two workers were very introverted and just replied with "yes" or "no" answers to our questions. However, they reacted strongly when we asked whether they were afraid of meeting their seniors in the café, explaining, "They don't have the right to discipline us here!" And the other remarked, "This is our private space, our factory should not discipline us here . . . after work, we have our personal liberty here." The juxtaposition of workers from throughout the hierarchy provides them with a subjective feeling that they are liberated from authority and from their workplace.

Workers go to the café predominantly to play Internet games. The gamers are usually the biggest spenders, of around 200 to 300 *yuan* a month, with some game addicts spending as much as 1,000 *yuan* a month, roughly a month's salary. According to one staff member of an Internet café, game addicts often do not go to the café after the middle of the month because by then their salary has been spent. But sometimes, if they cannot resist the temptation, they borrow money from their fellow workmates or relatives to visit the café. Such addicts spend more time in the café than the regular gamers, sometimes full days and nights when they have holidays. A café manager told us this is quite a common phenomenon, particularly when the workers quit their jobs. He explained that when they quit or were sacked and had some money remaining, they would have fun for a while before looking for another job. He also told us of one worker who spent three days in his café, playing continuously. Another worker dressed like a beggar because he spent a month playing games in his café. This beggar-like worker only took a short break each day for food and water. He guessed that he had not washed himself during that month. Whenever we asked the workers why they loved playing games, most of them pointed to the fun and satisfaction they found in the café. One game addict said:

> We [gamers] always meet at the same time; our biological clock has fixed the time. We love the atmosphere here, in between the virtual and the real. We have found satisfaction when we are playing the game. I may not be able to achieve in the real world, but I can play well in the virtual world.

Some workers told us that they meet players from all walks of life and they feel particularly pleased when they beat college students at games. Most of these game addicts are young migrant workers around age 20 who do not have any financial burdens. It is not the economic situation which pushes these workers to the cities; it is rather the coastal city life that has pulled them there. But instead of venturing into the city life in the Delta, these addicts seem to prefer the virtual life in the Internet café and in which they gain enjoyment and self-respect, particularly when they meet people from the upper strata of society. Once again, this is the result of the juxtaposition of the real and virtual.

Among the game addicts, it is not only the games which form the topic of communication, but also the language of the Internet or the game. We interviewed one game addict who spoke in a broken way and could barely make his ideas clear during the interview. At first, we thought that he might have a problem in communicating himself to others; but when we watched him playing a game, he became another person. He explained to us, in that form of Internet game language with which we were not familiar, the game which he was currently playing and the level he had attained. When he was explaining how to play the game, he was at the same time communicating with online friends. He told us that he tried to quit playing games for a couple of months but he found that he could not communicate with his workmates, and eventually he went back to the café again. His connection to the virtual prevails upon the connection in the real; this causes his communication difficulty in the face-to-face world.

Other workers indulge in the virtual world because they believe it provides them with fun and excitement. But do they really find relaxation and entertainment in the Internet café? Some told us that staying in the café was much more tiring than working in the factory. A factory supervisor told us that he could not stay online for more than an hour as his eyes would become very tired. When the workers are playing Internet games, they probably stay for hours online with their eyes focused on the monitor, while their fingers, which have been working for hours on the assembly line, work incessantly at the keyboard. Apparently, for these frequent visitors, their behavior seems more akin to a second shift than leisure time. The difference is that they are working on a cyber assembly line, rather than in a real one.[3] They believe they are relaxing, but their bodies suffer from fatigue after endless hours sitting in front of the machine. Some told us their time in the Internet café caused their work to suffer the next day. Leisure and entertainment have become another form of work. They earned points on the cyber assembly line, which, like money, are used to buy items necessary for extending their playing time. If they played more, they would improve. If they improved, they would be upgraded to a higher level of player, rather like being promoted at work. Through playing these Internet games, they operate in the cyberspace world of their dreams and aspirations to compensate for their frustration in the real world. They feel that they are free, have a certain measure of power and control and earn what they could not possibly have in the real world, even getting promoted to a higher level of skill. All this constructs an ideology of satisfaction—perhaps (following Lash 2002) this is the result of information domination.

CONCLUSION

When the new media technologies penetrated Chinese society in general and the migrant population in particular, it seemed much was promised.

New media technologies appeared to free migrant workers from both the spatial and temporal constraints of communication; they could extend their network from face-to-face to cyberspace, equipping them with strong social capital; they had access to an abundance of job market information, empowering them to negotiate with the capital. As our understanding of the influence of the new media technologies grows, we find ourselves having second thoughts about our previously positive view of this influence. We now find the new media technologies have also had a negative impact on the migrant population. The use of these new technologies may weaken the migrant worker's family ties, may exacerbate their loss in the host cities and may create illusions and lead them to another form of false consciousness. The description of the uses of the mobile phone and the Internet in this chapter is just the beginning of our observation of our recent research. We continue to gather new data and gauge developments framed by these first findings before making a more concrete theorization.

NOTES

1. http://stats/gov.cn/english/newsandcomingevents/t20090522_402560900. htm.
2. The study of migration discussed in this chapter was conducted by myself, doctoral students and colleagues at the Hong Kong Polytechnic University and the City University of Hong Kong. The chapter explores some of our early findings that have framed our continuing research program on the topic.
3. The idea of a cyber assembly line was coined by Chung-tai Cheng when we were conducting our fieldwork in 2007 in an Internet café in Shenzhen.

REFERENCES

Chan, Anita. 2001. *China's workers under assault: The exploitation of labor in a globalizing economy.* Armonk, NY: M.E. Sharpe.
Heidegger, Martin. 1992. *Parmenides,* trans. Andre Schuwer and Richard Rojcewicz. Bloomington: Indiana University Press.
Lash, Scott. 2002. *Critique of information.* London: Sage.
Law, Pui-lam, Yinni Peng. 2006. The use of mobile phones amongst migrant workers in southern China. In *New technologies in global societies,* ed. Law Pui-lam, Leopoldina Fortunati, and Shanhua Yang, 245–259. New Jersey: World Scientific Publisher.
———. 2007. Cellphones and the social lives of migrant workers in southern China. In *The social construction and usage of communication technologies: Asian and European experiences,* ed. Raul Pertierra, 126–142. Manila: University of the Philippines Press.
———. 2008. Mobile communication, mobile networks, and mobility: The case of migrant workers in southern China.' In *Handbook of mobile communication studies,* ed. James Katz, 55–64. Cambridge, MA: MIT Press.
Law, Pui-lam, Yinni Peng, and Cheng Chung-tai. 2007. Migrant workers, Internet cafe, and the Heterotopias: Paper presented in Beijing forum 'The Harmony of

Civilizations and Prosperity for All—Diversity in the Development of Human Civilization', Beijing, PRC, November 2–4, 2007.

Murphy, Rachel. 2002. *How migrant labor is changing rural China.* Cambridge: Cambridge University Press.

Ngan, Raymond, and Stephen Ma. 2008. The relationship of mobile telephony to job mobility in China's Pearl River Delta. *Knowledge, Technology and Policy* 21(2):55–63.

Pun, Ngai. 2005. *Made in China: Women factory workers in a global workplace.* Durham, NC: Duke University Press.

———. 2003. Subsumption or consumption? The phantom of consumer revolution in globalizing China. *Cultural Anthropology* 18(4):469–492.

14 Floating Workers and Mobile QQ
The Struggle in the Search for Roots

Chung-tai Cheng

INTRODUCTION

In this chapter I introduce the topic by firstly discussing the current status of mobile phone use in China today. This is then illustrated by an examination of the role and functions of mobile phone technology used for social relations among Chinese with an anecdote set in Dongguan city in Guangdong Province about two local people and "the mobile". Building on this example I will go on to suggest that the workers' situation in such a socio-technical landscape is totally different from that of local villagers because of their mobility and the characteristics of the charges for mobile phone services in China. The chapter continues by discussing how the Chinese Internet instant messaging, Mobile QQ, has combined with the mobile phone and how Mobile QQ constituted another form of social connection between workers in southern China. In conclusion, the chapter tries to show that such cyber connections may cause peasant-workers to succumb to a condition whereby they forget their former life-world and withdraw into their new floating or migrant world—one that could also be a virtual world.

The discussion in this chapter about social relations and communications in the lives of some peasant-workers is framed by Wellman's (2001) understanding of computer-supported social networks and of the socio-technical "geography of enablement and constraint" (Law and Bijker 1992). Focused on southern China, the chapter aims to contribute to the understanding of these peasant-workers' experiences of a different kind of social reality. In order to highlight the uniqueness and specificity of their particular situation, the chapter draws on a research study carried out by the author as well as on several other secondary research data. The primary research data are made up of five in-depth interviews and daily observations during a three-month stay in an industrial community, Daning village in Dongguan.[1]

The advancement of communication technology has allowed Chinese rural workers to benefit from new forms of social connection. Traditionally, people have socialized with each other by meeting in person, and social relations largely occurred through face-to-face interaction and group interaction, and relied upon physical presence (Thompson 1995). Today, however, relationship-building is not only understood as the outcome of the recognition of appearance or identity from face-to-face encounters, but is also considered possible through the establishment of other modes of connectivity. In particular, having contacts that are mediated via information and communication technologies (ICTs) is becoming one of the key elements of establishing social relationships. In this regard Wellman (2001) has suggested a tripartite typology comprised of groups, globalization and networked individualism to explain the establishment of social relationships in a framework of technological development of computer communications networks and flourishing social networks. He opined that now is a time for individuals and their networks, but not for groups, and furthermore that in networked societies boundaries are permeable, interactions are with diverse others, connections switch between multiple networks and hierarchies can be flattering and recursive (Wellman 2001). It is within this framework of networked and computer-mediated societies that I now consider how the lives of some of these peasant-workers in China have been transformed since the open door policy was implemented in 1978.

In only thirty years China is celebrating an economic miracle in fulfillment of Deng Xiaoping's vision, and peasant-workers, *nonming gong* or *minggong*, are one of the most important contributors to its impressive economic growth. During this time, thousands and thousands of workers have flooded from their villages into the towns, hoping to improve their quality of life and their prospects. However, after the peasant-workers born in the late 1960s and 1970s took to the road, a new generation of workers—those born and raised during the reform period—have been experiencing another journey. Although there was no complete break between these two generations of workers, the latter grew up in an environment surrounded by the media and hearing their predecessors' stories. This meant that the new generation of workers would come to cities with greater expectations and clearer images about how the journey would change their lives. This raises some important questions: What is the gap between workers' expectations and the reality? How is it revealed? How does it influence workers' perceptions of the future? In order to address these questions I now turn to some of the peasant-workers' anecdotes about their lives in this new environment of the city through which it is possible to learn some of the effects of the migration from the countryside to the city and how they have come to use ICTs in their daily lives to manage and mitigate the new social realities that they daily encounter.

MOBILE CONTACTS AS SOCIAL NETWORKING

One day, two locals, Ah Zhi and Ah Ren, discussed with me the problems of social order in their village. Ah Zhi said that he would only take his bank cards and mobile phone when going out to avoid being robbed. Unluckily, on occasion he forgot to bring his credit card while he was having lunch in a restaurant in another village. The waitress asked him to leave his mobile phone as security for his meal fee, but Ah Zhi refused and pledged his driving license only. A day later, Ah Zhi redeemed his license. When Ah Ren heard about this story, he reacted agitatedly: "How can that waitress request you to leave your mobile phone? It's ridiculous. Is she sick. . .Why didn't you give me a call? I could have come and paid for you!" "It isn't that good," Ah Zhi responded vaguely.

In the case above, the logic of the waitress's request is quite clear in the Chinese context, because the value of the mobile phone is higher than that of Ah Zhi's driving license. If necessary, she could sell the phone to recoup the money, whereas the driving license was useless to her. To Ah Zhi and Ah Ren, however, the value of the mobile phone is not only limited to its purchase value but also includes its practical value, e.g. when Ah Ren asked Ah Zhi why he did not call him. This means the mobile phone can function as a kind of spider's web connecting users in which social relations are no longer linear and limited by spatial distance but are like a spider's web with points where different threads join. If this is so, then how does the mobile phone influence Chinese social relations? Notably, if Chinese social relationships are understood as a "differentiated mode of association" which is established through social networks of personal relation with the self at the center and decreasing closeness as one moves out (Fei 1992), then what are the implications for Chinese society if the mobile phone becomes one of the media or mediators for establishing and maintaining social relationships? Every outsider to the village, including me when I visited it for my research, has to record personal details and apply for a temporary residence card or a working license to *Zibaohui*, or the Association of the Self-Protection for Local Social Order.[2] When Ah Ren knew I was going to stay in his village he called his own brother, who is the vice-chairperson of the Association, to settle my case directly. Although Ah Ren's brother was not willing to answer his calls and politely rejected his request, my application went smoothly.[3] The above anecdotes seem to have a similar logic: when you need help, look up your mobile phone and search for any possible contacts who can help you to solve your difficulties. It does not mean that the village does not have well-established and structured institutions or organizations to manage its society. What is surprising is that all characters in the stories tended to solve and settle their problems privately, which may suggest that relationships, or what we call *guanxi* in China, are far more important than formal and official but "inefficient" and "ineffective" channels. The meaning of *guanxi* is not limited to marital, familial and friendly relationships, it also carries a sense of "social connection" (M. Yang 1994). Many ordinary people think that mobile phones are facilitators to expand and maintain personal networks, but Yang's article suggests that the social implication of the advancement of communication technology in Chinese society is more complicated. There is no doubt that mobile phones can help maintain existing social relationships in expanded spatio-temporal contexts (Pertierra et al. 2002). People can make use of their mobile phone to keep in touch with whomever they like not only for strengthening their primary networks, but also for extending any possible *guanxi* outside kinship relations It is however, only once *guanxi* is established that one can ask a favor of another with the expectation that the debt incurred will be repaid sometime in the future (M. Yang 1994). In this context mobile phones can serve as buffers to avoid or relieve unwanted conflicts and embarrassment among Chinese people, such as by simply refusing to accept calls. From this perspective, the characteristic of social relationships as socially constructed is becoming more and more explicit. As Chinese social relationships are not the by-products of biological generation, a Confucian worldview, or any type of abstract "social structure" that works outside or above human subjects (Kipnis 1997), mobile phone technology can allow Chinese people to enjoy the situation in which they are maneuvering their social relationships. With the mobile phone, the practices of *guanxi* production and reproduction in Chinese society have become more and more autonomous, and social relationships between people are likely to become more purposeful and instrumental than before as the mobile phone becomes the means of delivering the social relations.

WORKERS: MIGRATING OR "FLOATING"?

> Once I met a new friend, Ah Quan, in our fieldwork site, but sadly he was going back to his hometown on the day we met. Although we spent just a few hours together, we wanted to stay in touch and exchange contacts. I asked him to give me his mobile number, but he refused and said, "My mobile number will be useless and abandoned once I get back home. Rather, I'll give you my QQ number so that we can meet on the Internet.

There are already some studies exploring the social influences of the rapid popularization of mobile phones among Chinese workers, especially those working in the Pearl River Delta of southern China. For example, Ke Yang (2008) has suggested that the practical value of mobile phones among migrant workers is to develop networks of *jianghu* friends. The idea of *jianghu* relations stresses the importance of friendship outside the primacy of kinship relations which, with the help of mobile phones, contributes to a kind of emotional support. It also has other advantages such as providing protection in group disputes and searching for work opportunities (K. Yang 2008). Law and Peng (2006) also suggested that mobile phones more closely connect kinsmen scattered far and wide among rural workers by not only allowing them to meet whenever and wherever they wish just by calling each other's phones, but also helping them to maintain closer contacts with family members who are physically far apart. Most importantly, they can prolong new social relationships in the workplace, developing friendships between workers from different parts of China even though they are not kinsmen and would normally consider each other as outsiders (Law and Peng 2006). In other words, because the mobile phone has contributed the most to the reconfiguration of the communicative sphere, its uses can strengthen and widen traditional *guanxi* and social networks (Fortunati et al. 2008). The social influences of mobile phones on both local people and workers does not seem to have any fundamental differences in that they allow mobile phone users to build social relationships, fulfilling a positive function in Chinese society.

The studies discussed in this section (K. Yang 2008; Law and Peng 2006) have in common the characteristic that they are viewing the "migration" of workers from their hometown to another place from a "static" image. By this is meant that workers tend to settle in one place either temporarily or permanently, and thus the use of mobile phones by workers does not have any fundamental difference compared with that of local villagers. It means they can establish and extend social networks in a new workplace with the help of mobile phones. The present study, however, suggests that the situation of the new generation of workers in Pearl River Delta is much more complicated, and is viewed from a more dynamic perspective. This is because the geographic mobility of the new generation of workers means that the ways they use mobile phones may be quite different. Because migration means moving from one place to another or a change of residence from one country to another in order to find a job or for other reasons, it is difficult to determine for how long a person's physical existence in a new place can be regarded as "migration". In order to understand this better the subjective identification of oneself with the place is the key point of investigation. Therefore, the concept of a migrant worker would be vague when workers' daily lives are not directly in touch with the societies in which they are living. As a matter of fact, many factories, companies and restaurants in Guangdong province provide meals and accommodation for their employees (Pun 2005). Theoretically, it is possible to meet one's basic needs only by staying in the workplace. A 28-year-old manager working in a restaurant shared this with me:

> I tried to stay in the workplace for a week. We need to work on shift. All shops would be closed if you work on night shift . . . Sometimes, I hang around during lunch break or go shopping, but I don't remember what the name of the market is. I just know how to go there . . . I don't care about all of this because I don't come from here or belong here.

More importantly, some recent statistics indicate that the phenomenon of workers changing their job is becoming more common in Pearl River

Delta.[4] In a study by Ngan and Ma (2008) they maintained that the better, more up-to-date and faster job information sent to migrant workers by their clansmen, friends and former co-workers over mobile phones acts as an additional catalyst to promote job change among migrant workers. These pieces of evidence reveal that some workers in southern China are moving or even "floating" between places, rather than "migrating". For example, the friend Ah Quan I introduced at the beginning of this section is only 22 years old, but he has been working outside his home town for four years. He changed his jobs three times: being a construction worker for half a year, a worker in a metal factory for one and half years, and a weaver for two years. When we met he had just quit his job and was ready to return back home to celebrate a relative's wedding ceremony. In addition to understanding the increasingly transient nature of the migrant workers, the question also remains regarding why Ah Quan chose to give me his QQ number but not his mobile phone one. The main reason for not using the mobile phone for this longer-distance contact is the practice of national roaming charges in China. Although charges for mobile services have dropped a lot in recent years, the abandonment of charges for interprovincial calls is still in dispute. To take China Mobile's M-zone services as an example, the charge for making calls in Guangdong Province both within the province and across provinces is 39 cents per minute. Although receiving regional calls is free of charge, calls from other provinces cost up to 29 cents per minute. In Guangxi Province, charges for making local calls in peak hours and in off-peak hours are 18 cents per minute and 10 cents per minute respectively, whereas the charge for making interprovincial calls is up to 60 cents per minute. Although receiving local calls is free of charge, calls from other provinces cost up to 40 cents per minute (Table 14.1)[5]

Table 14.1 Service Charges of China Mobile's M-zone Services

	Guangdong Province		*Guangxi Province*	
	Making calls	*Receiving calls*	*Making calls*	*Receiving calls*
Local intra-province	RMB 39 cents per minute	Free	RMB 18 cents per minute (peak hour)/ RMB 10 cents per minute (non-peak hour)	Free
National roaming (inter-province)		RMB 29 cents per minute	RMB 60 cents per minute	RMB 40 cents per minute

As Table 14.1 shows, the cost of using the Guangdong SIM card in the other provinces is higher than using the local SIM card to make or receive phone calls. Thus, if workers from the other provinces work in the Pearl River Delta, it is more economical and reasonable for them to change to a local mobile phone number once they go back home. Therefore, some of them only give their new numbers to those with whom they intend to keep in touch. The whole process relies solely on luck, however, because additionally no one knows whether anyone's mobile number is still valid or not. This phenomenon has an important implication: when workers are "floating" around, the social networks established in workplaces may not function well purely because of the loss of mobile contacts. If so, the primacy of kinship relations will then resume its role and power. In other words, mobile phones have the function of expanding and maintaining social networks, but it is not unlimited and depends on workers' own situations and should be considered alongside the socio-technical landscape in which it is rooted.

"FLOATING WORKERS" AND MOBILE QQ: THE STATE OF WITHDRAWAL INTO ONESELF

QQ, or Tencent QQ to give it its full title, is the most popular free instant messaging computer software in China. Apart from being an online chat program, QQ has also developed other sub-features such as QQ Groups as public membership discussion platforms, QQ-Zone as a personal blog, and QQ Games for online entertainment. According to its annual report for 2007, there were already 341 million active QQ accounts as of September 3, 2008, and it is estimated there are over 160 million QQ users in China. Because some users have two or more accounts, it is hard to know the exact number of users.[6] The total number of instant messaging users in China is 0.426 billion, in which the number of QQ users is up to 0.341 billion, representing 80.2% of the market share.[7] Because of its popularity, some scholars have already studied the social influences of Internet use on migrant workers in China. For example, Peng (2008) maintains that the Internet, especially online chats, enables migrant workers to maintain social ties established in their native provinces because it is an economical and convenient way to contact family members and friends.

Following on from the above studies, I now focus on another significant recent breakthrough: Mobile QQ—a combination of mobile phones and computer instant messaging technologies. In 2006, Tencent developed QQ to allow access across multiple platforms, including the PC and mobile phone. With Mobile QQ, mobile phone users can not only chat and meet with friends on the online discussion platform, but they can also share photos and songs, update their blogs (QQ-Zone), play Mobile QQ Games and enjoy movies or TV programs in Mobile QQ

Cinema. Most importantly, the charge for chatting online via Mobile QQ is cheaper than text messaging. For example, in Guangdong Province, China Mobile provides several SMS packages on M-Zone services such as RMB 19 for 130 messages, RMB 29 for 300 and RMB 39 for 500, but all are limited to the same network. Sending messages across the same network is 10 cents per message whereas sending between different company networks is 15 cents per message. Sending international messages costs RMB 1 per message. Chatting in Mobile QQ, however, is easier and more economical. Mobile phone users can choose two types of GPRS payments: monthly payment or daily payment. The former has several packets such as RMB 5 per month for 30M GPRS flows, RMB 20 per month for 150M and so on; the latter costs RMB 1 for 5M GPRS flows per day.[8] As a 22-year-old female security guard mentionedIt's more economical to chat in Mobile QQ. Each text message charges whereas you can talk freely in Mobile QQ. . . If you only chat with friends, the daily rate is good enough.

According to CNNIC (China Internet Network Information Center) there are already 0.4 billion mobile phone users in China. Of these, 12.6% had accessed the Internet on their mobile phone in the last six months.[9] Although the report does not contain information about use of the Internet on mobile phones by Chinese workers, the design of Mobile QQ has its own advantages that are expected to attract more and more mobile phone users to use it. First, QQ accounts can be used with Mobile QQ. As QQ numbers are computer bits located in cyberspace, physical places are no longer an obstacle in terms of establishing and maintaining social relationships. Even when workers go back to their hometown and change their mobile contacts, they can still keep their cyber networks by accessing Mobile QQ. It is also the reason why Ah Quan preferred giving me his QQ number rather than his mobile phone number. Second, mobile phones are empowered by this kind of technological innovation which can integrate with the Internet and then constitute a kind of "portable cyberspace". For example workers do not need to go to the Internet café because they can still contact friends in the Mobile QQ. Thus cyberspace is no longer constrained by computer portal and wire-lines but can really stretch its invisible hand into every corner of workers' daily lives. Third, the operation of the mobile phone is less complicated than that of the PC. Most of the Chinese-designed mobile phones have three types of input systems, namely phonetic input, stroke count input and pen-based input. As some workers are not well-trained in using Chinese phonetics or the computer, Mobile QQ can enable those low-educated or older workers to make use of the mobile phone chat room.

These facts, taken together with the situation of the new generation of workers in the Pearl River Delta, reveal an interesting but also possibly worrying picture that calls for further close consideration. While more and more workers in southern China are "floating" between places, it seems they cannot put down roots in their real world. As it happens, the advancement of communication technologies is likely to contribute another kind of connection and sense of belonging between workers. Therefore, the central question here is whether there are any differences between face-to-face and mediated communication? In face-to-face communication, social interaction must be based on certain kinds of social norms whereas mediated communication may weaken the effectiveness of a normative foundation in the real life world. Although these communications are not mutually exclusive, the key issue is how the differences between them are likely to reframe or even reconstruct workers' social relations and self-identification.

The Chinese notion of "self" is understood as being different from the rational, self-conscious and autonomous Western concept, as it is a "relational self" which tends to be socially or psychologically dependent on others; this "situation-centered" individual is tied closer to his/her relational world (Hsu 1981). It means that the essence of Chinese relationships is not about treating someone as an individual on equal terms, but rather one is considered according to the relationships between oneself and other persons. To understand the dynamic of Chinese relationship-building, one must participate in and learn from the experience of interacting and communicating with others in person. Only through this practice can one learn how to behave in accordance with the time and the place: e.g., if someone behaves socially or morally in an unacceptable manner, he/she would be rebuked and be urged to modify their behavior. During this process, one's self-identity can easily be established by the notion of behaving with *li* (propriety). To a large extent, meeting in person may even be considered as the representation of *li* itself. Therefore, face-to-face communication in Chinese society is essential for maintaining social relationships and confirming one's own social identity.

Unlike face-to-face communication, the normative foundation of the real world may be weakened and made looser in computer-mediated communication. Following Goffman (1973) and Miller (1995), because there is a lack of real life experience as a backdrop for the support of the separation of "front stage" and "back stage",[10] people can perform different roles and identities on the Internet. It may not be possible, or indeed may even be meaningless to care about one's own status and presentation in mediated communication via the Internet. Indeed it seems everyone in cyberspace is supposed to be an equal individual who can have certain autonomy to determine his/her own position. Thus the workers referred to in this present study may feel that they can stop and break their relationships in mediated communication freely, and in ways they would not in face-to-face real world contact. As a 22-year-old factory worker said,

> I can determine to be friends with anyone in QQ. If you meet a funny guy, then stay and chat with him. If not, you can just quit the conversation . . . Once a guy asked me some strange questions and spoke

rude words to me. I just blocked him and deleted his number from my contact list.

More importantly, workers may feel that they are searching for the opportunity to establish and expand "social" relationships throughout the process of mediated communication, but at the same time, due to the absence of a real situation, they can only use shallow titles to recognize others' identities and their relationships in the process. The result is that this kind of decontextualized interaction may foster a delusion about one's own existence among worker users. As a 28-year-old manager of a restaurant said,

> If my mobile phone was taken away, I don't think I could survive. You know, our dormitory is just a room with only four walls. It's damn boring. Every time when I am off work, I must log into Mobile QQ. Sometimes, I even fall asleep while I was still chatting with friends . . . If there is no one on QQ, I'll send messages to my friends and ask them to go online. From my point of view, I can throw away my husband but not the mobile.

A 22-year-old female security guard also revealed,

> Every night I chat with friends on Mobile QQ until I fall asleep. Somehow the mobile is the only thing that can help me to fall asleep.

The above conversations have an important implication: the new generation of workers is not only floating in between places as they move about geographically to different jobs, but it may also be enabling their withdrawal from their real world into cyberspace.

CONCLUSION

As there is already a great deal of literature about how workers are being exploited and even insulted by political and economic inequalities, this chapter tries instead to suggest that the situation of the second generation of workers in southern China is even more complicated if communication technology is taken into consideration. A common perception is that workers desirous of migrating from a village to a city, or from a poor environment to a relatively better environment, are viewed as "uncivilized" by local citizens. The chapter further suggests that this perception may distort and restrain the scope of investigation because the change of workers' identity and self-interpretation is a dynamic process. For example, a recent phenomenon is that workers in the Pearl River Delta are more mobile than before with some of them even floating around geographical locations in order to search for a place where they can settle down. Although they can establish face-to-face social relationships in workplaces with the help of communication technology, few can maintain these relationships because of floating between places. Mobile QQ as a technological communication innovation fills this gap, and some workers have started to adopt this kind of technology to escape the boredom and sadness of a dull and aimless working life. The paradox is, however, that the more they adopt and are addicted to cyber relationships, the more likely it is that they will feel lost and withdrawn from the real world in which they are living.

NOTES

1. This research program is supported by the Department of Applied Social Sciences at Hong Kong Polytechnic University. As I was born and grew up in Hong Kong, the perspectives and horizons of local Chinese and my own may be different. This different perspective and the cultural differences may be of benefit as they prompt some interesting questions. Special thanks to Dr. Pui-lam Law for his discussion and encouragement.
2. *Zibaohui* means literally the "association of self-protection", or Association of the Self-Protection for Local Social Order, which is a semi-official organization, mainly formed of local villagers. Because official and provincial police forces may not be capable of reaching every corners of the province, the *Zibaohui* performs some of their functions, being responsible for managing and monitoring several aspects of the village such as social order and public hygiene.
3. Ah Ren was quite unhappy about it because he felt that his brother had caused him to "lose face". Officially, I had to submit a copy of my ID card and to show any appointment letters from companies or factories in order to prove the reasons why I was going to stay in the village, so Ah Ren in fact had already helped me a lot.
4. Ngan and Ma (2008) studied the relationship of cell telephony and job mobility in China's Pearl River Delta by conducting questionnaire interviews with 655 migrant workers in three industrial towns in the Dongguan area. According to their survey, "Nearly all (79.24%) of the respondents had changed jobs at some point in their working life in southern China. Of these, 39.22% had changed jobs and employers in the year immediately before the interview period (i.e. April 2005 to March 2006). About half (50.58%) of the total number of respondents had left their jobs in Dongguan for a short period but then returned to work in the Pearl River Delta again" (56–57).
5. Data retrieved from Chinamobile's Web sites http://www.gd.chinamobile.com/m-zone/aboutus/zifei.shtml and http://www.gx.chinamobile.com/zifei/zifei/yl/dgdd/nn-dgdd2.html.
6. Data retrieved from http://www.tencent.com/zh-cn/content/ir/news/2008/attachments/20080319.pdf (dated October 13, 2008).
7. Data retrieved from http://www.analysys.com.cn/web2007/index.php?module=hysj&action=showone&id=5812 (dated October 13, 2008).
8. See http://www.gd.chinamobile.com/m-zone/aboutus/zifei.shtml and http://www.gd.chinamobile.com/m-zone/fee/conce/01_3336.shtml.
9. Data retrieved from http://tech.qq.com/zt/2008/cnnic21/ (dated October 13, 2008).

10. According to Goffman, people act out their lives "front stage" as others expect to see them and their true private feelings are kept "back stage" and shared only with those closer to them.

REFERENCES

Fei, Xiaotiong. 1992. *From the soil: The foundations of Chinese society*. Berkeley: University of California Press.

Fortunati, Leopoldina, Anna Maria Manganelli, Pui-lam Law, and Shanhua Yang. 2008. Beijing calling. . . Mobile communication in contemporary China. *Knowledge, Technology, and Policy* 21(1):19–27.

Goffman, Erving. 1973. *The presentation of self in everyday life*. Woodstock, NY: Overlook Press.

Hsu, Francis L. K. 1981. *Americans and Chinese: Passage to differences*. Honolulu: University of Hawaii Press.

Kipnis, Andrew. B. 1997. *Producing Guanxi: Sentiment, self, and subculture in a North China village*. Durham, NC and London: Duke University Press.

Law, John, and Wiebe Bijker. 1992. Postscript: Technology, stability and social theory. In *Shaping technology, building society: Studies in sociotechnical change*, ed. Wiebe Bijker and John Law, 290–307. Cambridge, MA: MIT Press.

Law, Patrick, and Yinni Peng. 2006. The use of mobile phones among migrant workers in Southern China. In *New technology in global societies*, ed. Pui-lam Law, Leopoldina Fortunati, and Shanhua Yang, 245–258. Singapore: World Scientific.

Miller, Hugh. 1995. The presentation of self in electronic life: Goffman on the Internet. Paper presented at Embodied Knowledge and Virtual Space Conference, Goldsmiths College, University of London, June 1995.

Ngan, Raymond, and Stephen Ma. 2008. The relationship of mobile telephony to job mobility in China's Pearl River Delta. *Knowledge, Technology, and Policy* 21(2):55–63.

Peng, Yinni. 2008. Internet use of migrant workers in the Pearl River Delta. *Knowledge, Technology, and Policy* 21(2):47–54.

Pertierra, Raul, Eduardo F. Ugarte, Alicia Pingol, Joel Hernandez, and Nikos L. Dacanay. 2002. *Txt-ing-selves: Cellphones and Philippe modernity*. Manila: De La Salle University Press.

Pun, Ngai. 2005. *Made in China: Women factory workers in a global workplace*. Durham, NC and Hong Kong: Duke University Press and Hong Kong University Press.

Thompson, John. B. 1995. *The media and modernity*. Oxford: Polity.

Wellman, Barry. 2001. Physical place and cyberplace: The rise of personalized networking. *International Journal of Urban and Regional Research* 25(2):227–252. http://www.itu.dk/people/sidsezimmermann/Litteratur/Wellman_2001%20-%20%20personalized%20networking.pdf.

Yang, Ke. 2008. A preliminary study on the use of mobile phones amongst migrant workers in Beijing. *Knowledge, Technology, and Policy* 21(2):65–72.

Yang, Mayfair M. 1994. *Gifts, favors, and banquets: The art of social relationships in China*. Ithaca, NY: Cornell University.

15 Community Connections and ICT

The Chinese Community in Prato, Italy and Melbourne, Australia—Networks, ICTs and Chinese Diaspora

Tom Denison and Graeme Johanson

INTRODUCTION

Diasporas, defined by Fullilove (2008) as "communities which live outside, but maintain links with, their homelands" (vii), appear to be inextricably entwined in the process of globalization and, as they become more organized, they are impacting relations both within and between host and home countries. Host countries are concerned with economic issues and questions of social cohesion, whereas home countries are increasingly attempting to engage with their diasporas, often in an attempt to gain economic advantage, by using their knowledge of markets, languages and business contacts (Fullilove 2008; Rindoks, Penninx and Rath 2006).

Advocates of the widespread dissemination of information and communication technologies (ICTs) are optimistic that communications networks may provide a bond between migrants, their places of origin and fellow migrants (Donner 2008). Barry Wellman, for example, articulates the potential benefits to such communities in using ICTs such as the Internet:

> [The Internet] is an excellent medium for supporting far-flung, intermittent, networked communities. E-mail transcends physical propinquity and mutual availability; email lists enable broadcasts to multiple community members; attachments and Web sites allow documents, pictures, and videos to be passed along; buddy lists and other awareness tools show who might be available for communication at any one time; and instant messaging means that simultaneous communication can happen online as well as face-to-face and by telephone. (2001, 2031)

Others, however, note that there could be broader potential consequences, for instance, in that modern ICTs could act "to elevate the importance of culture and downgrade that of geographical proximity" (O'Sullivan 2005). As Giddens points out, the threat to traditional structures could be far-reaching:

> The communications revolution and the spread of information technology are deeply bound up with globalizing processes . . . A world of instantaneous electronic communication, in which even those in the poorest regions are involved, shakes up local institutions and everyday patterns of life. (2000, 31)

Despite a spate of recent research into the importance of ICTs in relation to migrants and local community development (see, for example, Giorgiou 2002; Wilding 2006), the impact of these developments remains largely unexplored. This chapter sets out to at least partially address such questions, reporting on the results of two exploratory studies which investigate the use of ICTs by Chinese migrants in different settings. The first is an investigation into the use of Internet points by the Chinese community in Prato, Italy; the second is a more general investigation into the use of ICTs by Chinese migrants in Melbourne, Australia. Although starting from slightly different perspectives, both projects were designed to explore the role of ICTs in contributing to social cohesion and in maintaining social networks by exploring the use of ICTs by migrants and whether the use of ICTs by Chinese migrants has assisted in maintaining relationships, family and local identities, "homeland" cultural ties or in advancing business relationships or language learning.

The purpose of this research is to test the proposition that communications networks provide a bond between migrants, their places of origin and fellow migrants (Donner 2008).

The chapter reports on two distinct projects that contribute to our understanding. The next section will describe the work undertaken in relation to the Chinese-owned Internet points in China, whereas the subsequent section will describe the project involving Chinese migrants in Melbourne, Australia. These will then be tied together via a discussion section and lead in to the conclusion.

CHINESE INTERNET POINTS IN PRATO, ITALY

The first project centered on the use of Internet points by Chinese migrants in the Italian city of Prato. Prato supports a large textile industry and a ready-to-wear garment sector, characterized by a production model based on small businesses with a strong division of work processes, a structure which contributes to the dynamism and flexibility of the sector and provides many opportunities for entrepreneurial activity (Colombi 2002). As a consequence, Prato has been a popular destination for Chinese migrants since the 1990s. Bringing valuable skills and experience from home, Chinese migrants in Prato now manage some 2,400 small businesses, dominating the ready-to-wear garment sector and encompassing one quarter of the local textile industry (Pieraccini 2008).

Chinese migrants have proved a boon to the local economy, contributing to its general revitalization and opening up new opportunities by establishing economic links with China (Denison et al. 2009). However, there are well-documented negative aspects associated with the success of many Chinese businesses in relation to tax evasion, the use of illegal immigrants and sub-standard working conditions (Pieraccini 2008). Many commentators go further, representing the Chinese in Prato as alien outsiders unwilling to integrate into the local community (Di Castro and Vicziany 2007), and sensationalist stereotyping of the migrants in the media depicts protection rackets, serial economic crises and forced slave labor (Lee-Potter 2007).

Much of the debate on the problems associated with Chinese immigration centers on the degree to which the Chinese integrate with Italian society as opposed to forming closed communities. In a recent media report, Antonella Ceccagno (Ehlers 2006) observed that there is a mutual lack of trust and self-imposed separation between the Italian and Chinese communities, a sentiment echoed by the Mayor of Prato, Marco Romagnoli, who agreed that the Chinese are a "blessing" economically, yet concurrently "are a catastrophe for the community" (Ehlers 2006).

Throughout Italy, one focus of these concerns has been the proliferation of migrant-owned Internet points and telephone centers. These centers, often open twenty-four hours a day, have provided a focus for negative public opinion and have been the subject of a number of regional laws because of their perceived threat to public order and hygiene (Gandiglio 2007). Such laws have not yet been enacted in Prato, but their impact is being watched with interest by local authorities who are concerned that the significant amount of time spent by Chinese migrants in Internet points adds to the barriers to integration.

The research reported here collected data via a mix of survey questionnaires, completed by Internet point users, and interviews with managers. The intention was to draw out the issues with regard to the following questions:

1. In what settings do immigrants access ICTs and are the settings themselves important for realizing the benefits of ICTs?
2. Does the use of ICTs, or the settings in which they are used, have an impact on the ability of immigrants to integrate into the host communities?
3. What impact does the use of ICTs have on language skills, both with regard to the immigrants' first language and the language of their host country?

4. What are the implications for social capital of extended social networks maintained in this manner?

The six Internet points owned by Chinese in Prato are open twenty-four hours a day. Clustered in the main Chinese area around Via Pistoiese, they comprise two smaller centers, each with approximately 40 computers, three centers with banks of 100+ computers and one of 200 computers. These Internet points have been in operation for between three and eight years, and the managers interviewed had been working in them for between six months to eight years.

The user survey was based on a sample of convenience rather than one chosen on strict statistical principles. A total of seventy-five survey questionnaires were collected over a two week period in October 2007. Two young Chinese academics from Wenzhou were employed to help administer the questionnaire and lead the interviews. They took advantage of their language skills to tap into local social networks to find participants. Most of these were obtained by visiting the six Internet points and asking users to complete them as they used the service, but a small number were collected by approaching users in other venues, for example in telephone kiosks and shops.

A small group of eight workers in a local factory were also asked to complete the survey. These were approached because all had previously been users of the local Internet points, but no longer frequented them. These eight surveys were analyzed separately from those completed by current users.

FINDINGS

The Internet points in Prato are tightly focused on their core business of Internet access and online gaming. They try to differentiate themselves on price or by offering faster access or higher-quality graphics on their computers. Charges are kept low, typically 0.60 euros per hour, with discounts available to heavy users. They offer little or nothing in the way of other services. Only one offers a copy and print service and another offers technical support for the users' home computers. One point regularly issues sheets of job advertisements. The little food and drink that is available is usually sold through vending machines.

The responses to the questionnaire suggest strongly that most users conform to distinctive demographic characteristics. They are predominantly male. Some 70% are aged between 20 and 29, with only 12% aged 30 or over. They are also predominantly workers (63%), with students and the unemployed comprising 8% and 9% respectively. Ninety-two percent of users live and work in Prato.

Regular use was found to be common and prolonged. Thirty-one percent of respondents reported visiting the Internet point daily, with a further 39% coming twice or more a week. Just over 80% reported using the center for two or more hours per visit and of those, 55% spent over three hours per visit. As pointed out by the managers, however, levels and frequency of use varied according to the time of year. The interviews used in this study were collected in October and November of 2007, when the centers were about 50% occupied. According to the manager of one of the larger centers, up to 100 users might visit in a day, and up to 1,000 each week. Even so, they are much more heavily used during vacation times such as August and Christmas, with more users staying for longer periods.

By far the most popular use of the centers is for Internet chat (77%) and game playing (49%), typically MMORPGs (Massive Multiplay Online Role-Playing Games) that are played together with friends. The most popular other activities were: finding local information (11%), finding government information (8%), Internet telephone (8%) and e-mail (8%). Those who surfed the Web to any extent visited a variety of Web sites with the most common being Chinese language search engines and music download sites.

Seeking relaxation after long hours of work was a common motivation, with some adding that they came to "make life more colorful". In addition, potentially self-improving uses of the Internet points included: making new friends through the Internet (63%), making contacts for work (20%) and reading work advertisements (11%).

A large proportion of users (67%) had Internet access at home or work but continued to use the centers because the speed of access at home was too slow, and because they prefer to share the social environment with friends. The socializing function of the points is extremely important in two senses. One is the sharing attributed to the physical meeting place of the point itself, the other the social activities that the visitors use the Internet for. Eighty-one percent of visitors came with one or more friends, not alone.

Of the eight ex-users who completed the questionnaire based on their past use of these services, seven were between the ages of 25 and 34, and one was over 35. These responses were not processed with the other responses due to their limited number and the fact that they were based on past rather than current use, but overall their responses were similar to those of current users. Of particular interest were their comments that although they had used Internet points when they were younger as a way to relax and relieve the stresses of long hours of work, as they started families and became more established in their communities they used the centers less and less. They believed that this was a common pattern.

This picture contrasts strongly with that of the other Internet points in Prato, whether they are managed by Italians or south Asians. The scale of these businesses and the range of services offered are completely different. Their opening hours approximate to standard business hours, with most offering between eight and twenty computers. They provide international telephone calls, fax, photocopying, money transfer and courier services. Although there is some social engagement involved in their use, most users come alone and visit for periods of only twenty to thirty minutes. Communication via e-mail and chat is common, with the focus being on information-seeking activities and business or professional uses. Game playing is minimal.

It is clear that the Chinese-owned Internet points in Prato are primarily entertainment centers, places to go and relax with friends after work, rather than cogs in a communication network. They are similar to the centers that can be found in any large city in Australia, and no doubt many other places in the world, located in urban entertainment areas. Having said that, it is also clear that a lot of communication goes on in and through them, and although most of it is in Chinese, those that can speak Italian reasonably make use of them for multilingual communication.

CHINESE MIGRANTS IN MELBOURNE, AUSTRALIA

The second project, although seeking to explore similar themes, took quite a different approach in terms of both its immediate subject and its methodology. It examined levels of social cohesion of Chinese and Italian immigrants and the role of ICTs in shaping their networks. Groups of Chinese and Italian migrants were interviewed in Melbourne, Australia, to discover varying patterns and processes of communications for multilateral social relationships, for family and local identities, for acquiring local knowledge, for retaining "homeland" cultural ties, for advancing business relationships, for language learning and for socio-economic progress. The Italian and Chinese communities were chosen because, although both have integrated into mainstream society, the Italians are well-established and the Chinese are more recent.

Migrants come to Melbourne for many reasons, but the most common are to provide necessary skills for employment (55%) and for family reunion (26%). Victoria has a total population of 4,932,422, representing 18% of Australia's population. Most migrants in Victoria were born in England, but the second-largest source of migrants is Italy, and the fifth-largest is China. Migration accounts for half of the current population growth in Victoria. In the period from 2000 to 2005, 40% of migrants to Australia came from Asia (Arunachalam and Wulff 2008). Arunachalam and Wulff (2008, 45) detail some demographic features that demonstrate the potential for integration of Chinese migrants in Melbourne:

> They are gradually integrating into the host society while maintaining their cultural identity. Over time, they are as likely to own a home as the average Australian; they are even more likely to live in those areas of suburbs where social, economic and cultural activities are likely to be

concentrated (inner city area and middle area suburbs); these areas are likely to have above average house prices. The dominant family types among the Chinese people in Melbourne are similar to the ones among the Australian-born, although more among the Chinese are likely to share living quarters with unrelated others (reflecting the overseas students from China). Finally, the Chinese people are marginally more likely than other [migrant]s to take up the Australian citizenship.

The project used grounded theory to provide understanding of both immigrant and host communities and the basis of their communicative relationships. Semi-structured interviews were adopted as the best means to obtain detailed understanding of the migrants' views. Potential interview questions were derived from prior readings of literature and by speaking with individuals who had a good general knowledge or first-hand experience of the topics to be researched. The issues identified initially were consolidated and re-arranged into a sequential interview schedule of questions and statements, which were then piloted on early participants. After data collection, the content of the audio-recorded views of participants were classified into common themes and concerns, then summarized to characterize group views, with any significant deviations noted. Data analysis was iterative, taking place even during the collection of views via the interviews, and more intensely in the categorization period after the interviews. Overall themes which emerged for each group were then compared with each other.

Although twelve Chinese and twelve Italian migrants were interviewed for this project, this chapter focuses on the results of the interviews with Chinese migrants.

Findings

Interviewees used the full gamut of mobile technologies for communicating, as well as for study, travel, immigration, banking and accessing business information. Share-trading was also very common. Other uses mentioned included access to Chinese comics, news about recent earthquakes in China, online lectures, gaming, Chinese recipes, e-commerce, housing prices, directories and entertainment.

Eight important themes emerged from the interviews:

1. *Family.* Family links seem stronger among the Chinese than any other links. Even physical distance does not remove a firmly held family obligation to stay in regular touch. Frequent family visits to and from China are commonplace,whether they are ostensibly for work or family reasons.
2. *Generations.* Young and adult Chinese (under 60) were adept at using ICTs and exploited their potential extensively, whereas those in their

sixties rely more on television, radio, fax and newspapers. Chinese in Melbourne are early and comprehensive adaptors of all forms of ICTs, and will try any new products and services; they are very IT savvy, especially with regards to costs and value-for-money.

3. *Immediacy of phones and messaging.* There is a strong preference among most interviewees for use of the phone, especially mobile phones. One interviewee even called phone talk "face-to-face communication", implying that it is as close as anyone could come to real contact. The younger Chinese in Melbourne (those in their twenties) strongly favored instant messaging and chat, with QQ being the most popular service, followed by MSN Messenger.

4. *Language.* A knowledge of Chinese is required to access online information; however, Chinese as a language does not lend itself to easy keyboard construction. Obstacles to language construction reinforce the ease-of-use and portability offered by the mobile phone. One highly educated interviewee could not type Chinese very quickly: "Four lines [of e-mail] would take me half an hour," because her dialect is different from Mandarin.

5. *Frequency of contact.* Contact with China was very frequent, with at least eight of the twelve interviewees making "daily" contact. Without daily contact, one interviewee said that he would be "desolated". Occasionally another interviewee would e-mail his parents, but "mum is very bad at computer stuff. If you care about the relationship, you use the common language." In this instance, the "common language" means the common medium, the mobile phone.

6. *Physical distance.* Distance from the homeland does not seem to affect the pressing desire for family connection. Time zones (two hours difference) are not seen as a problem for phone calls. Access to ICTs is hard in parts of rural China, but there is no difficulty in cities. Trips back to China and from China to Melbourne are very regular, as mentioned.

7. *Hierarchy of technologies.* There was evidence of an unarticulated hierarchy of ICTs. Mobiles (whether by voice or SMS) were used frequently for the least formal, most friendly occasions, along with chat and instant messaging. E-mail was used for more formal communications, for recording business, children's development (photographs) or obtaining official information, for instance about immigration rules. The Internet was used for a wide range of purposes, including leisure and game-playing activities, and for work. The most precious information content was archived, primarily on CDs, but also printed on paper.

8. *Maintenance of original cultural identity and relations within the host country.* Evidence about cultural identity and social outreach (making moves to integrate with the adopted mainstream culture) is ambivalent. Few interviewees were freely forthcoming on this point,

not having reflected on it before. There is no doubt that the frequent communication pattern indulged in by all interviewees bolsters the intensity of their relationships with their Chinese family and friends. Although it is possible to identify a cultural enthusiasm for comprehensive adoption of ICTs, it is more difficult to show that the content of communications represents a shared culture.

It seems clear that the use of ICTs facilitates a feeling of belonging and well-being of Chinese migrants in Australia. There is evidence for integration, but it is partial. It may be that the wrong question is being asked here—maybe "integration" is not so relevant in a globalized world, where freedom of movement is easier and virtual connections are guaranteed. ICTs are used extensively by Chinese migrants in Melbourne to support the maintenance of their community identity and memory, but so are they in the country of origin. ICTs help Chinese migrants maintain social, business and cultural links to China, and vice versa.

DISCUSSION: COHESION OR SEPARATISM?

Tantalizingly, the research has revealed seeds of constructive social networks and examples of communications engagement, but it has also raised questions relating to other features of cohesion, such as the complex set of relationships between place and both personal and group identity.

In Prato, there is persistent anecdotal evidence pointing to the negatives, suggesting that some users are addicted and that, even for those who are not, the long hours spent in the points may have serious consequences for family life and other activities such as school attendance. Yet there is also consistent evidence pointing to the positives: providing a motivation for improved literacy skills, facilitating access to Chinese culture, and maintaining relationships with family and friends abroad.

As they stand, the Chinese-owned Internet points of Prato are a stark reminder of the differences between the Chinese and Italian communities, differences that are manifested in socio-economic backgrounds, employment types, personal motivations and habits of public relaxation. These points clearly act to bond members of the Chinese community and to reinforce those contacts. On the other hand, it is hard to conclude that the Chinese Internet points provide a bridge or are a means of integration into Prato society. However, that does not mean that they do not serve an important need within their own community, nor does it mean that they do not have important benefits for the community as a whole. There is potential for general social benefits, but for the moment they seem to function as islands in the mainstream. Use of the Internet per se is both common and important to both communities, but it serves different purposes for Chinese and Italians. Further study is needed, to explore the relationships between social capital and the social networks that not only sustain Chinese immigration to Prato, but also bind these communities together while making the most of the growing international networks.

The Chinese in Italy represent a recently-established community which retains a strong association with its home country. In drawing comparisons between the findings of these two studies, it is necessary to remember that the two countries follow contrasting policy settings in that, for example, unlike Australia, Italy encourages assimilation rather than multiculturalism (Kerkyasharian 2008), and favors unskilled rather than skilled migration. Further, the Chinese diaspora in Europe is well-connected internally, regardless of its external relationships to local groups, and the need to bridge to traditional local cultural experiences does not seem to be strong among them. Often there is little initial intention to put down local roots, but rather to roam Europe "as a chessboard" seeking any work prospects (Smith 2004). This may not be the same as in Australia.

As noted by Maya-Jariego and Armitage (2007, 743) in their investigation of migrant networks in Spain:

> Modern communications have weakened the traditional relationship between physical setting and social space. Enabling participation in multiple communities simultaneously, physical presence is no longer necessary or a guarantee for participation . . . While as individuals we give meaning to our realities across a complexity of communities, our relationships are continuously situated in time and space.

It cannot be denied, however, that technology represents an important point of contact in a global society and the lives of the young, and a more considered approach to the services provided by Internet points is warranted. Ceccagno (2004), as part of her study of second generation Chinese immigrants, reported evidence of both positives and negatives in the use of local Internet points. On the positive side, she found that use of the centers provided the opportunity for young Chinese to embrace Chinese culture (primarily but not exclusively music); a motivation for learning written Chinese; the opportunity to make and/or stay in touch with networks of friends in China or elsewhere; and, for those that are more competent in Italian, the opportunity to make more links with the Italian community.

Ceccagno (2004) believes that younger Chinese are more educated than their parents, and in general more competent with Italian and familiar with the overall culture. Whereas their facility with Italian and new technologies potentially opens up new prospects for them, there is a lack of social mobility, and in looking for points of reference and contact in the country in which they live they meet many impediments. Thus, younger Chinese can be frustrated and disengaged in their attempts to integrate, and the time spent isolated from the broader community only reinforces these problems (Ceccagno 2004).

For the Chinese, there is no doubt that the frequent communication pattern indulged in by all interviewees bolsters the intensity of their relationships with their family and friends. Other research, such as Wilding's (2006) study of a number of migrant groups in western Australia, have reported that although most migrants willingly embrace the ease of communications afforded by ICTs, there remain some that "regretted the ways in which new ICTs reduced their capacity to sustain a sense of distance . . . precisely because they found their home country socially or culturally stifling or their kin dominating and difficult" (135–136). However, migrants are free to adopt several identities and can vary their identities for a variety of reasons, depending on social and economic settings, to fulfill a patriotic need or a cultural connection, or to strengthen emotional bonds to family or friends. Migrant identity is not necessarily the primary identity adopted in all situations. Thanks to worldwide ICTs, migrant identity can be shaped and personalized by the migrant to fit in with freshly adopted lifestyles, which are not dependent on geographical place (Pung 2008).

There is a potential for the supportive intervention of ICTs. It is essential to accept that "for the foreseeable future, ethnic identity is likely to continue to be an ubiquitous, hotly debated, and politicized social-psychological phenomena" (Kim 2006, 295). Because identity is founded on so many facets of life—nationality, region and community of origin, ethnicity, language, lifestyle, organizational affiliations, age, social class and spiritual beliefs and practices, it is very fluid (Chuang 2004).

CONCLUSION

Thus, evidence from these studies about cultural identity (bonding) and social outreach (making bridges to connect with the adopted culture) is ambivalent. Although it is possible to identify a cultural enthusiasm for comprehensive adoption of ICTs, it is more difficult to show that the content of communications represents a shared culture.

Individuals belong to multiple communities, and migrants are no exception. Migrant groups in diasporas are like "World Wide Webs" in their own right, "With dense interlocking, often electronic strands spanning the globe and binding different individuals" (Fullilove 2008, ix). As Giorgiou (2002) argues, diasporas have always relied on networks, and "the construction of shared imagination, images and sounds have always been key elements of sustaining community" (Giorgiou 2002, 3); the appropriation of ICTs to suit the circumstances of their everyday life is to be expected.

Independent of the act of physical migration, global virtual networks remain intact, firmly in the control of the migrants, functioning like portable safety nets to shore up culture and identity (as required) against the rigors that accompany physical migration, their own renewed development and possible future threats to inherited culture. The characteristics of these networks, and the role of ICTs in shaping them, require further study.

REFERENCES

Arunachalam, Dharmalingam, and Maryann Wulff. 2008. Chinese settlement in Melbourne. *Around the Globe* 4(3):42–45.

Ceccagno, Antonella. 2004. *Giovani migranti cinese: La seconde generazione a Prato.* Milan: Franco Angeli.

Chuang, Rueyling. 2004. Theoretical perspectives: Fluidity and complexity of cultural and ethnic identity. In *Communicating ethnic and cultural identity,* ed. Mary Fong and Rueyling Chuang. New York: Rowman and Littlefield.

Colombi, Massimo, G. 2002. Migranti e imprenditori: Una ricerca sull'imprenditoria cinese a Prato. In *L'imprendtorialità cinese nel distretto industriale di Prato,* ed. Massimo, G. Colombi. Milan: Franco Angeli.

Denison, Tom, Arunachalam Dharmalingam, Graeme Johanson, and Russell Smyth. 2009. The Chinese community in Prato. In *Living outside the walls: The Chinese in Prato, Italy,* ed. Rebecca French, Graeme Johanson, and Russell Smyth. Cambridge: Cambridge Scholars Publishing.

Di Castro, Andrea A., and Marika Vicziany. 2007. *Chinese dragons in Prato: Italian-Chinese community relations in a small European town.* Melbourne: Monash Asia Institute.

Donner, Jonathan. 2008. Research approaches to mobile use in the developing world: A review of the literature. *The Information Society* 24(3):140–159.

Ehlers, Fiona. 2006. The new wave of globalization: Made in Italy at Chinese prices. *Spiegel Online International,* September 7. www.spiegel.de/international/spiegel/0,1518,435703,00.html (accessed November 2008).

Fullilove, Michael. 2008. World wide webs: Diasporas and the international system. Working Paper 22, Lowy Institute. www.lowyinstitute.org/Publication.asp?pid=753 (accessed November 2008).

Gandiglio, R. 2007. *Diritto and Diritti.* Turin, Italy: Divisione Commercio della Città di Torino.

Giddens, Anthony. 2000. *The third way: The renewal of social democracy.* Malden, MA: Polity Press.

Georgiou, Myria. 2002. Diasporic communities on-line: A bottom up experience of transnationalism. London School of Economics and Political Science. www.lse.ac.uk/collections/EMTEL/Minorities/papers/hommesmigrations.doc (accessed November 2008).

Kerkyasharian, Stepan. 2008. Defending multiculturalism. *Around the Globe* 4(3):26–27.

Kim, Young Yun. 2006. From ethnic to interethnic: The case for identity adaptation and transformation. *Journal of Language and Social Psychology* 25(3):283–300.

Lee-Potter, Adam. 2007. Designer labels' sweatshop scandal. *U.K. Sunday Mirror,* December 2. www.sundaymirror.co.uk/news/sunday/2007/12/02/designer-labels-sweatshop-scandal-98487-20191613/ (accessed November 2008).

Maya-Jariego, Isidro, and Neil Armitage. 2007. Multiple senses of community in migration and commuting. *International Sociology* 22(6):743–766.

Pieraccini, Silvia. 2008. *L'assedio cinese: Il distretto parallelo del pronto moda di Prato.* Milan: Il Sole 24 Ore Libri.

Pung, Alice, ed. 2008. *Growing up Asian in Australia.* Melbourne: Black, Inc.

O'Sullivan, John. 2005. The real British disease. *The New Criterion* 24:16–23.

Rindoks, Aimee, Rinus Penninx, and Jan Rath. 2006. *Gaining from migration: What works in networks? Examining economically related benefits accrued from greater economic linkages, migration processes and diasporas.* IMISCOE (International Migration, Integration, and Social Cohesion in Europe) Working Paper 13, prepared for the OECD (Organisation for Economic Co-operation and Development) Development Centre.

Smith, T. 2004. Crisis in Tuscany's Chinatown. BBC News, news.bbc.co.uk/2/hi/Europe/3500285.stm (accessed November 2008).

Wellman, Barry. 2001. Computer networks as social networks. *Science* 293 (September 14): 2031–2034.

Wilding, Raelene. 2006. "Virtual" intimacies? Families communicating across transnational contexts. *Global Networks* 6(2):125–142.

16 Imagining China
Online Expatriates as "Bridge Bloggers" on the Chinese Internet

David Kurt Herold

INTRODUCTION

In 2006, the *China Daily* reported over 150,000 foreigners legally working in China, with many more in-country long-term as frequent business travelers, students, spouses, etc. (Wang 2006). Among these over 500,000 long-term residents in China, there are many who write about their experiences in online blogs or similar Web sites.[1] Motivations for these blogs differ widely, and among the bloggers are language students wishing to share their knowledge of Chinese; lawyers working in China; journalists who want to write about stories they cannot publish with their employers; individuals or groups who vent their anger and frustration with life in China, etc.

Despite the large number of expatriates blogging about China, they have so far been ignored by the academic community. There are many studies of Chinese cyberspace and its relationship with real events in China (e.g. Goldsmith and Wu 2006; Guo 2007; Herold 2008, 2009; MacKinnon 2008), but nothing on the opinions and attitudes of expatriate bloggers in China, despite their demonstrable influence on the opinions of, for example, foreign journalists publishing news stories in European and American media outlets.

Being a foreigner in China sets an individual apart. Writing about the comparatively more cosmopolitan and globalized environment in Singapore, Chai (2006) showed how U.S. expatriates in Singapore form a separate "privileged" group as a result of their own and other people's reactions to "whiteness". This same reaction to being visibly different affects expatriates living in China and constitutes a major influence on both an individual expatriate's perception of identity, as well as his/her identification with a group. As Tai (2009) showed, the effect of being confronted with difference is not only applicable to Western expatriates living in Asia, but also Japanese encountering these expatriates, with both groups defining themselves through their interaction with each other (for identity formation through group interactions, see also Ailon-Souday and Kunda 2003). Both expatriates and Chinese define their identities through their interactions with each other and in opposition to each other.

Abdelal et al. (2006) refined this process of identification by arguing that the identity of a group (and of individuals within a group) is a social construct defined by their shared opinions (content), as well as the interpretation of their differences to "the other" (contestation). In a similar manner Janssens, Cappellen and Zanoni (2006) demonstrated how female expatriates utilize interactions with male expatriates to create "effective professional identities," (Janssens et al. 2006, 143) while setting themselves apart from both their male colleagues as well as the population of their host country.

The groupings that emerge out of the repeated interactions with perceived "others" form communities of shared identities, which define themselves—and are defined by "others"—by the differences to those "others", as well as the common understandings they share. Fernandez, Mutabazi and Pierre (2006) argued that this process of "othering" is particularly extreme for expatriates in China, as the history of the Western encounter with China and its attendant long process of (self-)orientalization of China means that "interpretations of Chinese reality rest both on the traditional 'Them and Us' duality and on the notion of 'another world' perceived to be radically unknown, that is physically and mentally inaccessible" (Fernandez, Mutabazi and Pierre 2006, 14).

As will be shown below, expatriates living in China assign themselves a position in-between two perceived extremes: Chinese people living in China, and European and Americans living in their home countries. They claim to possess superior knowledge to the knowledge either has about the other, and frequently assume the role of intermediaries, attempting to build "bridges" of understanding, for which many of them utilize the expressive affordances of the Internet.

This chapter wants to provide a first step towards a better understanding of expatriates in China and expatriate bloggers in particular by providing an overview of the different types of Web sites run by expatriates, which is followed by a brief description of the community of expatriates as a whole, including an outline of some of their opinions about China and the Chinese people. The chapter concludes with a number of questions and pointers for further research into expatriates in China.

EXPATRIATES IN CHINESE CYBERSPACE

Many of the foreigners living in China share their experiences, both professional and personal ones, with others online. Some do so anonymously, while others link their online presence to their business presence in China. These sites are linked through link networks as well as mutual referrals and comments on each other's sites, and cover a wide range of different topics, approaches, styles, etc. The following overview over some of the more frequently visited sites is meant to provide merely a first glance at some of the variety on offer, but is by no means exhaustive. The overview allows for further study of expatriates in China by the interested reader, offering several points of departure for additional research.

Listings and Pointers

The first category of sites to be introduced here is a testimony to the organization and mutual awareness of the expatriate community in online China. The people who create and maintain English-language Web sites about China do not only focus on their own Web site or blog, but are aware of and peruse other expatriate Web sites as well and compare themselves to those other sites.

For the past few years, the "China Analyst" Web site has run annual competitions for the "Best English language China blog" in six different categories every year during December. Internet users can nominate and vote for the best blogs in different categories. While the vote itself is not too serious, the voting process usually results in discussions of the relative merits of different sites in the comment sections of those Web sites open to comments, which promote connections within the community.

The "China Blog List" is an attempt to provide a listing of all English-language blogs that deal with China, and was started by an expatriate who has been blogging about features of the Chinese language for the past few years on the Web Site "Sinoplace". Although it is impossible for any one site to cover all the sites with posts on China, the site's listings of 10 Random, 10 Newest and 10 Hottest Blogs offer interesting starting points for a closer look at expatriate blogs in China.

The last two sites to be mentioned in this section do not claim to offer exhaustive lists or even an objective overview. The list at "Alltop" is hand-picked by the site owners and provides a useful RSS-feed-like listing of the most recent posts on a long list of China-related blogs and Web sites. Without even following the links, the different post headers shown allow for a glimpse into the variety available on sites posting about China.

Last, but not least, is the "HaoHao Report" which functions like the American site Technorati. The Chinese word *hao* means "good", and the site allows its users to post links to individual posts on China that they thought were good. Other users can follow the links to the listed posts and add their votes agreeing or disagreeing with the original poster. This enables visitors to the site not only to find Web sites or individual posts addressing issues in China, but also to gauge the reaction of the wider expatriate community to the posts.

Explanatory Efforts

This next section is the one that Cooper (2009) identified as bridge-building blogs or Web sites. The Web sites in this category are engaged in efforts to bridge the gap between Chinese Web sites and the wider world by translating, explaining and commenting on events and developments in China. They attempt to present objective accounts of what is happening in China, but are often over-emphasizing scandals or over-representing problems in the sensationalist manner of tabloid newspapers—something Cooper mentions in her article as well.

The first type of Web sites in this category are sites that translate news stories from the Chinese media and Chinese cyberspace into English for the benefit of their international readers, who they hope will understand China better as a result of their efforts (e.g. Fauna 2009a).

"Chinasmack" is a Web site with a very active group of members who frequently engage in long discussions in the comment sections after translated articles. The site shows a preference for translating the more lurid stories from Chinese cyberspace, involving sex, drugs, violence, etc. They cover a wide range of topics, but are all stories that have attracted a high level of attention from Chinese netizens. Recent examples of topics include a Chinese man's complaint about the size of his girlfriend's breasts (Fauna 2009b), Chinese reactions to the screening of the movie *2012* in China (Fauna 2009c) or a government official's offensive statement to reporters (Jessie 2009).

"ChinaHush" is another Web site specializing in the translation of often sensationalist news items and online forum discussions and was created by a Chinese American who calls himself Key. It came into being during 2009 and has therefore less people following it than the older Chinasmack, but it often provides stories that Chinasmack does not cover such as stories about the lack of rural sex education in China (CC 2009), or U.S. president Obama's visit to China (Key 2009b). The two sites, used jointly, cover most of the hottest topics on the Chinese Internet and serve as a fascinating resource for studies of debates about them among online expatriates.

Key emphasizes that his Web site is intended as a way "to create a community with a common interest in China" (2009a). Despite the relative youth of the site, he has been very successful with this intention, as, for example, the more than 1,000 fans of the site on Facebook show, or the more than 200 people following each new posting on Twitter.

Two Web sites with a longer history are "Danwei" and "ESWN" which have existed for a number of years and are well known throughout the Chinese expatriate Internet. Danwei provides media-related translations of Chinese news stories, official announcements and Internet postings (e.g. Liu 2009; Martinsen 2009). Roland Soong's ESWN translates political stories from China and often collates them with Western media stories, attempting to provide a neutral repository of "hot" debates in Chinese cyberspace (Soong 2009a, 2009b).

Neither of the two Web sites is known for its active discussions, but their stories are widely read and referred to. Roland Soong's translations are often quoted by other Web sites, news stories in Europe and America and by academics writing about the Chinese Internet, because of their accuracy and the clear separation of the translations from the occasional comments by the site owner.

A very different approach to the one adopted by the translation sites is taken by two sites run by academics interested in the Chinese Internet and journalism in China in general. Both of these sites offer translations and summaries of events in China, but are more activist, as they want to critically evaluate developments in modern China. "China Digital Times" (CDT) is run by the Berkeley China Internet Project and covers "China's social and political transition and its emerging role in the world" (China Digital Times 2009). Their summaries deal with political stories from a variety of sources and include links to those sources with discussions of the stories and events. Recent examples include a report on the Chinese Blogger Conference 2009 (Beach 2009) and a discussion about online freedom and censorship in China (Qiang 2009).

"RConversation" by Rebecca MacKinnon is a personal blog by a former bureau chief of CNN's Beijing office who is interested in journalism and censorship in China. Her blog, and the "Global Voices Online" site which she co-founded in 2004, are useful sources of information on the efforts of the Chinese government to control the Chinese Internet. The posts tend to exaggerate the importance of a "free press" and to frame all stories as fights between an oppressive Communist government and courageous journalists and citizens fighting for the establishment of "true" democracy in China. They are nevertheless valuable sources for emerging government legislation, or the mood among Chinese academics and activis

EXAMPLES OF EXPAT-SPECIFIC INTERESTS

The first group of sites in this category are sites that were created by their owners to allow them to vent some of their frustrations with China and life in China. They are often collections of posts retelling events that happened to the posters and that they found difficult to deal with. Outsiders coming across these sites often complain that they are too negative or even prejudiced about China, which is usually countered by pointing to the owners' long-time exposure to China and by claiming that their online venting about "bad China days" is healthier than taking their mood out on the people around them.

FOARP ("Fear of a Red Planet") is one of the people venting their frustrations online, while keeping his offline identity a secret. He blogs about his experiences in China and comments on news stories about China in the U.S. media. As his name indicates, he is generally critical of China and its government (see FOARP 2009).

The most famous of this type of Web site were two by-now-defunct Web sites run by groups of expatriates with the explicit intention of venting their anger about "bad China days". The first site, "TalkTalkChina", ran from June 2005 through December 2006 and was succeeded by "Sinocidal", which was online from December 2006 until April 2008. Their archived postings offer a fascinating source of anecdotes about the life of expatriates working in China.

The next two groups of sites addressing their owners' specialized interests are an extension of the employment of the owners. Firstly, there are a number of sites on which journalists blog, who use their blogs as additional outlets for stories their employers have no space for or do not want to publish. The China correspondents change from time to time but the practice has been well-enough established to be continued by new journalists arriving to replace an outgoing China reporter; see for example the switchover from Richard Spencer (Spencer 2009) to Peter Foster (Foster 2009) for the U.K.'s *Telegraph* newspaper.

These journalistic blogs allow the reporters more leeway in their comments and in the scope of stories they cover than a news corporation can. The best-known blog in this category is the one run by John Pomfret, who used to be the *Washington Post's* Beijing bureau chief, and is currently the editor of the *Post's* Outlook section. His blog "Pomfret's China" covers a variety of stories about China, the Chinese Internet, expatriate postings on China, etc., and invites comments and criticism from his readers—see also the "China Rises" blog by Tim Johnson, who left China in early 2009 after six years of reporting from Beijing for the McClatchy newspapers.

The second group to be mentioned in this context is corporate lawyers commenting on Chinese laws, regulations and court cases on their blogs. They use their blogs and Web sites to discuss the general developments on China's legal scene, which provides an interesting insight into the business practices of foreign companies in China.

The biggest of these, and the blog repeatedly receiving awards as best law blog on China, is "China Law Blog". The blog is written by two international lawyers, Dan Harris, based in the United States, and Steve Dickinson, based in China. The blog's focus is business law in China, but it addresses a whole range of legal issues facing international companies in China, from Joint Venture Laws to Immigration Laws or negotiation strategies for foreign companies in China (Harris 2009).

Stan Abrams uses his blog "China Hearsay" to talk about intellectual property regulations and international trade issues (see Abrams 2009). William Moss's blog "Imagethief" deals with public relations issues of companies and government offices in China (see Moss 2009).

A final type of Web sites with special interests are those run by students or teachers in China, who use their blogs to share their knowledge of the Chinese language. On the "Chinese Language Blog" the authors share anecdotes and discoveries with their readers, such as the Chinese phrase for "being ashamed" (Lopin 2009), while Sinosplice offers systematically arranged language lessons in Chinese for different levels, as well as insights into Chinese grammar, e.g. the aspect system in Chinese (Pasden 2009).

SHARED IMAGINATIONS

An Online Community of Expatriates

Expatriates writing and commenting about China on blogs and Web sites offer a wide variety of contents. It should be noted, however, that they are not merely a large number of people who happen to produce similar Internet presences, but that they do form a networked community of people who frequently interact with each other.

Posts on one of the sites that attract the interest of other expatriates are often cross-posted to other sites, with discussions in the comment sections expanding the basis for discussions to include other sites and their communities of posters and commentators in debates. One recent example is the debate around racism in China, for which Steve from "Fool's Mountain" decided to put together a post summarizing the story of a young Chinese woman who was abused on the Chinese Internet because of her mixed racial heritage (Steve 2009), and included links to other English and Chinese postings on the topic. The discussion in the comment section was joined by many observers and participants in the online expatriate community in China, e.g. Fauna from Chinasmack and FOARP. The basis of the discussion expanded to include not only Steve's summary, but also articles on the affair published on Chinasmack, ChinaHush, James Fallows's blog (Fallows 2009), the Shanghaiist (Chow 2009) and many other sources. While this is only one example, it still demonstrates how close the links are that exist between the different sites posting about events in China.

Beyond the formal closeness of the networked community, however, members of the expatriate community in China also have fairly similar opinions about China. While their opinions are not identical, their disagreements are often reminiscent of the fights of an old couple who already know that they disagree about a certain topic, but fight about it again anyway (there are several examples of this type of behavior in the discussion mentioned above).

As the archives of TalkTalkChina and Sinocidal show, even the topics under discussion keep recurring, if a Web site runs for too long a period of time. The image, or to use Benedict Anderson's term, "imagination" (1991), of China shared by the expatriates posting and commenting about China and life in China includes a number of specific features that keep emerging on different Web sites across the years. Although the image is not complete, and there are variations in the way different sites approach topics or evaluate nuances, certain memes can be found across the expatriate Internet that are shared by the community as a whole.

Imagining China and Its Citizens

This section will attempt to provide the reader with an impression of expatriate views and attitudes through the discussion of the two underlying beliefs shared by expatriate Web sites and blogs. While in no way exhaustive, such a description will serve to allow the reader to assess both the general attitude of the online expatriate community to China as their "host country" and the perception the community has of itself as not belonging to China or the non-Chinese wThe first basic belief is that the Chinese state is both misunderstood and misrepresented by and in the non-Chinese world. Sites like Chinasmack, ChinaHush, ESWN, etc. were created to address this misunderstanding of China through the translation of Chinese language materials into English to make their content available to non-Chinese people.

Among the "mistakes" frequently mentioned by expatriates are several that—to the expatriates—explain why other countries have so many problems with China, e.g. the assumption, particularly in the U.S., that China is a "Communist" country similar to the Stalinist USSR. Particularly the legal and economic blogs, e.g. China Law Blog, China Hearsay, Imagethief, etc., often feel the need to explain that China is one of the most Capitalist countries in the world and that Capitalism is not softened in China by a strict adherence to laws, protection of the less affluent or a social security system.

Another fiction of European and American news stories often mentioned as such is the supposedly strict and oppressive control over the Chinese people exercised by the Communist government, which ordinary Chinese people are trying to rebel against. Stories on Chinasmack and on China-Hush, as well as discussions on Imagethief, Fool's Mountain or the Peking Duck, frequently demonstrate the lack of control the Chinese government has over the people in China and the often fervent nationalism of most Chinese (see e.g. Page 2009).

The second basic belief that is shared by expatriates living in China and often expressed is about the image many non-Chinese have of Chinese people and how wrong this image usually is. Foreign businesses are frequently advised to forget most of what they have heard about Chinese people by business or law related blogs. Imagethief and China Hearsay have covered many stories of foreign companies being cheated by their Chinese business partners, or getting into trouble with individual Chinese bureaucrats, etc. because they misinterpreted the behavior and statements of their Chinese partners. Sites like Chinasmack, ChinaHush, the Peking Duck, the Lost Laowai and more have published many stories about foreigners with misperceptions about "Chinese people" being mistreated by individual Chinese and stories of events in China that do not fit non-Chinese assumptions about the behavior of Chinese people.

Common misconceptions many non-Chinese are presented as having include the assumption that Chinese people are either mild-mannered Confucians or Buddhists, living in harmony with nature, etc. or that they are the exact opposite, being oppressed Communists, living in Stalin-era concrete jungles. However, many expatriate sites such as the Culture section on the Lost Leeway blog show through examples from daily life that the behavior of most Chinese today is not compatible with such images, but includes elements from all of these sources mixed with a very large amount of Capitalism.

ESWN frequently translates Chinese Internet posts in which the authors explain that for contemporary Chinese people money is far more important than anything else. Modern Chinese are presented as being less interested in abstract concepts pursued by Europeans and Americans, e.g. "Human Rights", "Democracy", etc., and far more interested in acquiring ever more and ever more expensive luxury goods, e.g. large TVs, cars, houses, etc. Imagethief and China Hearsay have repeatedly warned foreign companies that Chinese companies are mostly interested in earning money, and that they will ignore the possibility of long-term benefits for a short-term financial gain.

In general, expatriate sites seem to agree that Europeans and Americans tend to have too positive an image of Chinese people, while having too negative an image of the Chinese state and the country as a whole (see e.g. DJ 2009). For expatriates, this applies almost in reverse. Many expatriates like China and enjoy their lives in-country, but are frequently frustrated in their contact with Chinese people (e.g. Joel 2008). As a simple Google search reveals, many expatriates complain of having "bad China days" whenever Chinese people "are being too Chinese"; see for example the answers posted on the forum of "The Beijinger" (*Beijinger* 2009).

Expatriates in China do not portray themselves as being in tune with the society in which they live, but they also do not think of themselves as still part of the societies they came from. Instead they claim to belong to a special group, expatriates, who are between China and the non-Chinese world, agreeing with neither side and misunderstood by both. While they frequently assert that they possess superior knowledge about China (and about their "home" societies) and therefore have the right to publish blogs, Web sites or even books about their experiences, there are few attempts to "convert" outsiders to their point of view—the "outsider" just wouldn't be able to understand (see e.g. DeWoskin 2005; or Kitto 2009).

QUESTIONS AND POINTERS

Expatriates in China who use the Internet to engage with other expatriates form a coherent and identifiable group. As a group, they are well-connected, and frequently engage each other in debates about topics related to China. A large part of their discussions about China serve to imagine China as a misunderstood country with a population of misunderstood people.

The self-distancing of the group from their countries of origin, as "China experts", is matched by the distance expatriate postings show as existing between the group and Chinese people. Expatriates in China identify themselves as knowing more about China than other Europeans or Americans in the countries of their origin, but also as being different from—and often superior to—the Chinese people they live with.

While this chapter can serve as an introduction to the community of online expatriates in China, there are still many questions left unanswered and deserving of further study. The links provided here can serve as a first stepping stone for researchers to dig deeper and to understand this fascinating community better. Two areas of further study that this chapter did not touch upon, but that are very necessary, are the nationality and race of the expatriates within the community and the relationship between the expatriate community and Chinese netizens who stumble across their Web sites.

With very few exceptions, the Web sites and blogs discussed in this chapter are run by people of English-speaking European or American origin, of mostly Caucasian descent. Considering the vast differences in experiences in China by people from different language backgrounds, national origins and races, any communities created by them online would almost certainly not function in the same way as the community presented here. A comparison of the online habits and the community formation among expatriates from different backgrounds in China could yield some highly valuable insights not only into national differences but also into the complex relationship between the different "foreigners" living in China and the Chinese people.

The relationship between Chinese netizens and the expat community is another area that urgently requires attention. Some Chinese have risen to prominence in the expatriate community, for example Key, the Chinese American owner of ChinaHush, or Fauna, the Chinese owner of Chinasmack. Other Chinese netizens, however, rarely take part in the discussions on expatriate Web sites or blogs, and if they do, they are easily identifiable as Chinese "outsiders" in this expatriate community.

The members of the community seem to think that they are better informed, better educated, better—than most Chinese people. It would therefore be highly interesting to study the interactions between Chinese commentators and the expatriate community to look at Chinese perceptions of this community and its beliefs about China and the Chinese people. Be it the bridge-blogging posts on ESWN or Chinasmack or the "bad China days" across many other sites, the reaction of Chinese people to the posts would add another dimension to the understanding of expatriate communities in China and serve as an additional verification of the efficacy of the communities' efforts in building bridges of under-standing between the Chinese and the non-Chinese world.

NOTES

1. Web sites and blogs will be referred to by name with a listing of the Internet addresses at the end of the chapter.

WEB SITES AND BLOGS

Alltop	http://china.alltop.com/
China Analyst	http://www.chinalyst.net/
China Blog List	http://www.chinabloglist.org/
China Bounder	http://chinabounder.blogspot.com/
China Digital Times	http://chinadigitaltimes.net/
China Hearsay	http://www.chinahearsay.com/
ChinaHush	http://www.chinahush.com/
China Law Blog	http://www.chinalawblog.com/
China Rises	http://washingtonbureau.typepad.com/china/
Chinasmack	http://www.chinasmack/
Chinese Language Blog	http://chineselanguageblog.com/
Danwei	http://www.danwei.org/
ESWN (East South West North)	http://www.zonaeuropa.com/
FOARP	http://foarp.blogspot.com/
Fool's Mountain	(Fool's Mountain Blogging for China) http://blog.foolsmountain.com/
Global Voices Online	http://globalvoicesonline.org/
HaoHao Report	http://www.haohaoreport.com/
Imagethief	http://news.imagethief.com/blogs/china/default.aspx
Internet Archive	http://www.archive.org/web/web.php
James Fallows	http://jamesfallows.theatlantic.com/
Lost Laowai	http://www.lostlaowai.com/
Peking Duck	http://www.pekingduck.org/
Pomfret's China	http://newsweek.washingtonpost.com/postglobal/pomfretschina/
RConversation	http://rconversation.blogs.com/rconversation/
Shanghaiist	http://www.shanghaiist.com/
Sinocidal—	A record of the site can still be found at: http://web.archive.org/web/*/http://www.sinocidal.com
Sinosplice	http://www.sinosplice.com/
TalkTalkChina—	A record of the site can still be found at: http://web.archive.org/web/*/http://talktalkchina.com
Technorati	http://technorati.com/
The Beijinger	http://www.thebeijinger.com/

REFERENCES

Abdelal, Rawi, Yoshiko M. Herrera, Alastair Iain Johnston, and Rose McDermott. 2006. Identity as a variable. *Perspectives on Politics* 4(4):695–711.

Abrams, Stan. 2009. IP as loan collateral: Another China innovation policy. *China Hearsay*, November 20. http://www.chinahearsay.com/ip-as-loan-collateral-another-china-innovation-policy/.

Ailon-Souday, Galit, and Gideon Kunda. 2003. The local selves of global workers: The social construction of national identity in the face of organizational globalization. *Organization Studies* 24(7):1073–1096.

Anderson, Benedict R. 1991. *Imagined communities: Reflections on the origin and spread of nationalism*, rev. ed. London and New York: Verso.

Beach, Sophie. 2009. Chinese BloggerCon 2009: Micro power from the mouth of a cave. *China Digital Times*, November 14. http://chinadigitaltimes.net/2009/11/chinese-bloggercon-2009-micro-power-from-the-mouth-of-a-cave/.

Beijinger. 2009. Bad China days: How often do you have them? General Discussion Forum. *The Beijinger*, October 30. http://www.thebeijinger.com/forum/2009/10/30/Bad-China-days-How-often-do-you-have-them.

CC [pseud.]. 2009. The blank slate of sexual education in the Chinese countryside. *ChinaHush*, November 19. http://www.chinahush.com/2009/11/19/the-blank-slate-of-sexual-education-in-the-chinese-countryside/#more3553.

Chai, Rosemary. 2006. White U.S. expatriate professionals in Singapore: Desiring to be cosmopolitans. *Intercultural Communication Studies* 15(2): 118–132.

China Digital Times. 2009. About China Digital Times. http://chinadigitaltimes.net/about/.

Chow, Elaine. 2009. Lou Jing talks to Netease about Oriental Angel, growing up black. *Shanghaiist*, September 16. http://shanghaiist.com/2009/09/16/lou_jing_talks_to_netease_about_ori.php.

Cooper, Marta. 2009. China: Bridging the gap? Interviewing bridge bloggers. *Global Voices Online*, October 30. http://globalvoicesonline.org/2009/10/30/china-bridging-the-gap-interviewing-bridge-bloggers-in-china/.

DeWoskin, Rachel. 2005. *Foreign babes in Beijing: Behind the scenes of a new China*. New York: W. W. Norton and Company.

DJ [pseud.]. 2009. Ah, that tricky Chinese propaganda machine, how devious it is to deceive the foreign media! *Fool's Mountain: Blogging for China*, November 17. http://blog.foolsmountain.com/2009/11/17/ah-that-tricky-chinese-propaganda-machine-how-devious-it-is-to-deceive-the-foreign-media/.

Fallows, James. 2009. Festival of updates #8: Chinese/US attitudes on race, flu. *The Atlantic*, September 7. http://jamesfallows.theatlantic.com/archives/2009/09/festival_of_updates_8_race_iss.php.

Fauna [pseud.]. 2009a. About Chinasmack. *Chinasmack*. http://www.chinasmack.com/about/.

———. 2009b. Dumping girlfriend because her breasts are too small? *Chinasmack*, November 18. http://www.chinasmack.com/stories/dumping-girlfriend-breasts-too-small/.

———. 2009c. 2012 movie insults China, Chinese netizen reactions. *Chinasmack*, November 20. http://www.chinasmack.com/stories/2012-movie-insults-china-chinese-netizen-reactions/.

Fernandez, Bernard, Evalde Mutabazi, and Philippe Pierre. 2006. International executives, identity strategies and mobility in France and China. *Asia Pacific Business Review* 12(1):1–24.

FOARP [pseud.]. 2009. Sentencing in Shishou. *Fear of a Red Planet blog*, October 18. http://www.foarp.blogspot.com/2009/10/cross-posted-at-gongshangfa-remember.html.

Foster, Peter. 2009. Obama-Hu: A dialogue of the deaf. *Telegraph Blogs*, November 17. http://blogs.telegraph.co.uk/news/peterfoster/100017085/obama-hu-a-dialogue-of-the-deaf/.

Goldsmith, Jack, and Tim Wu. 2006. *Who controls the Internet? Illusions of a borderless world.* Oxford and New York: Oxford University Press.

Guo, Liang. 2007. *Surveying Internet usage and its impact in seven Chinese cities.* Beijing: Center for Social Development, Chinese Academy of Social Sciences.

Harris, Dan. 2009. China negotiating strategy: An expert's perspective. *China Law Blog,* November 19. http://www.chinalawblog.com/2009/11/china_negotiating.html.

Herold, David Kurt. 2008. Development of a civic society online? Internet vigilantism and state control in Chinese cyberspace. *Asia Journal of Global Studies* 2(1):26–3.

———. 2009. Cultural politics and political culture of Web 2.0 in Asia. *Knowledge, Technology, and Policy* 22(2):89–94.

Janssens, Maddy, Tineke Cappellen, and Patrizia Zanoni. 2006. Successful female expatriates as agents: Positioning oneself through gender, hierarchy, and culture. *Journal of World Business* 41(2):133–148.

Jessie [pseud.]. 2009. Only CCP members can question government officials. *Chinasmack,* November 15. http://www.chinasmack.com/stories/only-communist-party-members-can-question-government-officials/.

Joel [pseud.]. 2008. (Letter) Lots of us want to love and respect China, but right now China isn't helping. *Fool's Mountain: Blogging for China,* August 23. http://blog.foolsmountain.com/2008/08/23/lots-of-us-want-to-love-and-respect-china-but-right-now-china-isnt-helping/.

Key [pseud.]. 2009a. About ChinaHush. *ChinaHush.* http://www.chinahush.com/about/.

———. 2009b. President Obama took questions from fake Chinese students at town hall meeting. *ChinaHush,* November 16. http://www.chinahush.com/2009/11/16/chinese-students-who-president-obama-took-questions-from-are-fake-students/#more-3519.

Kitto, Mark. 2009. *Chasing China: How I went to China in search of a fortune and found a life.* New York: Skyhorse Publishing.

Liu, Alice Xin. 2009. In New York: Contemporary heroes from China's music scene. *Danwei,* November 17. http://www.danwei.org/music/contemporary_heroes_from_the_u.php.

Lopin, Aaron. 2009. You should be ashamed, very ashamed. *Chinese Language Blog,* August 9. http://chineselanguageblog.com/bid/10215/you-should-be-ashamed-very-ashamed.

MacKinnon, Rebecca. 2008. Flatter world and thicker walls? Blogs, censorship and civic discourse in China. *Public Choice* 134(1–2):31–46.

Martinsen, Joel. 2009. A new look for the Beijing Morning Post. *Danwei,* November 20. http://www.danwei.org/front_page_of_the_day/a_new_look_for_the_beijing_mor.php

Moss, Will. 2009. Sail a river of moonshine at Guizhou's expo pavilion. *Imagethief,* October 28. http://news.imagethief.com/blogs/china/archive/2009/10/28/sail-a-river-of-moonshine-at-guizhou-s-expo-pavilion.aspx.

Page, Bob. 2009. Are online relationships between China and the US boiling over? Rednecks against Red Guards? *Mercury Brief,* October 31. http://www.mercurybrief.com/2009/10/red-guards-and-rednecks/.

Pasden, John. 2009. Aspect, not tense. *Sinosplice,* November 19. http://www.sinosplice.com/life/archives/2009/11/19/aspect-not-tense.

Qiang, Xiao. 2009. Tao Weishuo defends his statement: "All Chinese people have Internet freedom". *China Digital Times,* November 17. http://chinadigitaltimes.net/2009/11/tao-weishuo-%e9%99%b6%e9%9f%a1%e7%83%81-defends-his-statement-all-chinese-people-have-Internet-freedom/.

Soong, Roland. 2009a. A blogger reflects on blogging. *EastSouthWestNorth (ESWN),* http://www.zonaeuropa.com/20090921_1.htm.

————. 2009b. The Guanxian County cyber cafe shutdown. *EastSouthWestNorth (ESWN)*, October 19. http://www.zonaeuropa.com/20091019_2.htm.

Spencer, Richard. 2009. Will China get unhappier? *Telegraph Blogs*, April 3. http://blogs.telegraph.co.uk/news/richardspencer/9372993/Will_China_get_unhappier/.

Steve [pseud.]. 2009. Lou Jing: Racism gone wild? *Fool's Mountain: Blogging for China*, October 21. http://blog.foolsmountain.com/2009/10/21/lou-jing-racism-gone-wild/.

Tai, Eika. 2009. Multiethnic Japan and Nihonjin. In *Japan's minorities: The illusion of homogeneity*, ed. M. Weiner. Abingdon, VA: Routledge.

Wang, Renmin. 2006. Number of foreigners working in China soars. *China Daily*, April 4. http://english.peopledaily.com.cn/200604/04/eng20060404_255781.html.

Contributors

Maria Bortoluzzi is a lecturer of English language in the Faculty of Education (University of Udine, Italy). Her research interests and publications are in the fields of critical discourse studies, multimodality, teaching and learning English as a second language. Her latest research work deals with multimodal analysis within a framework of critical discourse studies; she is interested in the developments of verbal and non-verbal communication in ICT discourses. In the Faculty of Education, she teaches English in multimedia and technology courses (undergraduate), teacher training courses (post-graduate) and collaborates with the Ph.D. program of the Faculty of Modern Languages and Literatures.

Cigdem Bozdag, is research assistant in the research project "Communicative connectivity of ethnic minorities: The integrative and segregative potential of digital media for diasporas" at the ZeMKI (Center for Media, Communication and Information Research) at the University of Bremen, Germany. She is working on a dissertation on diasporic cultures and the appropriation of diasporic Web pages in the Moroccan and Turkish Diasporas in Germany.

Chung-tai Cheng has a Ph.D. from the Sociology Department of Peking University, with his undergraduate degree in Social Policy and Administration at The Hong Kong Polytechnic University. He has been researching and publishing in journals and books on the normative implications of ICT uses in China, especially in relationship to self-identification and social relationships among Chinese rural workers, on which his ethnographic study focuses. His presentation "Imagined performativity: the great virtue of cyberspace in contemporary Chinese workers' social lives" won the Best Presentation Award in the COST Action 298 2009 Conference: The Good, the Bad, and the Challenging.

Cristiano Codagnone joined IPTS as a senior scientist in October 2009. He has a degree in economics from Bocconi University (Italy), and a Ph.D. in sociology from New York University (1995). In the 1990s he was visiting fellow at the Institute of Sociology of the Russian Academy of Science and contract professor at the Institute of Sociology of Bocconi University. He also completed his post-doctoral study at Utrecht University in 1996. From 2000 to 2003, he was a faculty member at the Department of Social and Political Studies, Milan State University, and from 2003 to 2004 he served as project officer in the United Nations' "eGovernment for Development" program. Since 2004, he has carried out several studies on eHealth, eInclusion and eGovernment for the European Commission.

Tom Denison is a research associate with the Centre for Community Networking Research (www.ccnr.net) in the faculty of Information Technology at Monash University. He conducts research within the fields of social and community informatics, specializing in research relating to the effective use of information and communications technologies (ICTs) by communities and their members. Particular foci of his research include: ICTs as a form of mediated communication and the consequences of such use, for example by migrants and migrant groups; the use of ICTs by non-profit organizations and cultural institutions (libraries in particular); the role of social networks; and barriers to the adoption of ICTs.

Clifton Evers is a lecturer and researcher at the University of Nottingham, Ningbo, China. He researches gender, media, sport and cultural studies, with a particular interest on social change. His recent book Notes for a young surfer (2010) is a cutting-edge account of experiences of young men. Clifton is a chief investigator on a number of research grants and his research makes use of creative methodologies that can capture a broad range of media experiences—cognitive, physical, spatial and social. He is an editor-in-chief of Altitude: An e-Journal of Emerging Humanities Work and Kurungabaa: A Journal of Literature, History and Ideas From the Sea.

Leopoldina Fortunati (Editor) teaches Sociology of Communication and Culture at the University of Udine, Department of Human Sciences, and is the director of the Ph.D. program on Multimedia Communication. She has conducted several research projects in the field of gender studies, cultural processes and communication and information technologies. She represents Italy in the COST Domain Committee "Individuals, Societies, Cultures, and Health" and is very active at a European level. She is the co-chair of the International Association "The Society for the Social Study of Mobile Communication" (SSSMC). Her works have been published in eleven languages: Bulgarian, Chinese, English, French, German, Italian, Japanese, Korean, Russian, Slovenian, Spanish.

David Garbin is a research fellow at CRONEM (Centre for Centre for Research on Nationalism, Ethnicity and Multiculturalism, University of

Surrey). He is conducting fieldwork among diasporic French-speaking Africans, mainly Congolese migrants in London and Atlanta, and British Asians. He is also undertaking fieldwork in the Democratic Republic of the Congo (Kinshasa and Nkamba) as part of his ethnography of the Kimbanguist church, one of the largest African independent churches. In collaboration with Gareth Millington (Roehampton University) he is also exploring issues of post-colonialism, youth identities and urban/ spatial marginalization in post-riot France. His research interests include transnational religion, African and South Asian diasporas, migration, globalization, diasporic processes, popular culture and the politics of identity and ethnicity in urban settings.

Gerard Goggin is Professor of Media and Communications at the University of Sydney. He is author of New technologies and the media (2012), Global mobile media (2011), Cell phone culture (2006), and, with Christopher Newell, Disability in Australia (2005) and Digital disability (2003). Gerard is also editor of various collections on mobiles and Internet, including Mobile Technology and Place (2011), Mobile technologies: From telecommunications to media (2009), Internationalizing Internet studies (2009), Mobile phone cultures (2008) and Virtual nation: The Internet in Australia (2004). This chapter draws upon a national study of youth and mobile media in Australia, funded by the Australian Research Council.

Lelia Green is Professor of Communications in the School of Communications and Arts at Edith Cowan University (ECU), Perth, Western Australia, and a chief investigator in the Australian Research Council's Centre of Excellence for Creative Industries and Innovation (CCI). From 2006–2009, Lelia was a member of the International Advisory Panel for the E.U. Kids Online I Project, co-funded by the European Community's Safer Internet Programme and the London School of Economics and Political Science. She has retained her involvement with the E.U. Kids Online Project and, with Professors John Hartley and Catharine Lumby, also of the CCI, conducted parallel research in Australia between 2010 and 2012. The author of Communication, technology and society (2002), Lelia's most recent book is The Internet: An introduction to new media (2010). The work reported here benefited from Australian Research Council funding, and an ECU Early Career Researcher grant to Dr. Nahid Kabir.

Heike Mónika Greschke studied Social Work, Sociology and Social Anthropology in Koblenz and Bielefeld (Germany) and in Seville (Spain). She is currently leader of the junior research group "climate worlds" at Bielefeld Graduate School in History and Sociology, where she had received her doctoral degree in 2008. She is author of the book *Is There*

. *a Home in Cyberspace? The Internet in Migrants' Everyday Life and the Emergence of Global Communities* (forthcoming). She is a former research fellow at the Centre of Globalization and Global Governance, University of Hamburg, and at the Department of Educational Science, Bielefeld University; her main research interests are ethnomethodology and ethnography applied to the study of globalization phenomena in everyday life.

Andreas Hepp is Professor of Communications at the ZeMKI (Center for Media, Communication and Information Research) at the University of Bremen, Germany. He is author and co-author of six books as well as co-editor of a number of further books, including Connectivity, network and flow: Conceptualising contemporary communications (2008, with Friedrich Krotz, Shaun Moors and Carsten Winter) and Media events in a global age (2009, with Nick Couldry and Friedrich Krotz). One of his main research areas is media, migration and diasporas.

David Kurt Herold taught and researched in China for over nine years before joining the Hong Kong Polytechnic University in 2007 as a lecturer of Sociology. His research is focused on the use of ICTs by humans. In particular, he studies the Chinese Internet, encounters between Chinese and non-Chinese online, the impact of the Internet on offline society and online education. His recent publications include Cultural politics and political culture of Web 2.0 in Asia (2009); Mediating media studies: Stimulating critical awareness in a virtual environment (2010), Imperfect use? ICT provisions and human decisions (2010); On-line society in China (2011, edited volume).

Heather Horst is a socio-cultural anthropologist at the University of California Humanities Research Institute, University of California, Irvine, interested in the relationship between place, space and new media. She is the co-author of The cell phone: An anthropology of communication (Horst and Miller 2006), Hanging out, messing around and geeking out: Kids living and learning with digital media (Ito et. al. 2009) and is working on a manuscript that examines the relationship between communities, networks and society among youth and families living in the Silicon Valley region. She is currently carrying out research, funded by the John D. and Catherine T. MacArthur Foundation, on the relationship between learning ecologies, networks and pathways in Chicago and New York.

Graeme Johanson, Associate Professor, is Director of the Centre for Community Networking Research (www.ccnr.net) in the Faculty of Information Technology at Monash University. His research has covered Internet communities, social networks, civil society organizations, capture of indigenous cultural heritage in perpetuity by use of information and communications technologies, organizational knowledge sharing and academic and public libraries. His consulting work has taken him to the two World Summits on the Information Society, sponsored by the U.N. in 2003 in Geneva and in Tunis in 2005, as a member of the official Australian delegation. Several current research projects involve a comparative study of the use of information and communications technologies by migrant communities.

Nahid Kabir is a Senior Research Fellow in the International Centre for Muslim and non-Muslim Understanding at the Hawke Research Institute, University of South Australia. Dr Kabir was a Visiting Fellow (August 2009–July 2011) in the Islam in the West Program at the Center for Middle Eastern Studies, Harvard University, USA. Nahid Kabir is a consultant for UNESCO for Human and Social Sciences. She is also an advisor of Islamopedia which is a web-based resource on contemporary Islamic thinking (managed by the Islam in the West Program, Center for Middle Eastern Studies, Harvard University, USA). Nahid Kabir is the author of *Muslims in Australia: Immigration, Race Relations and Cultural History*, London: Routledge 2005. Her second book is entitled, *Young British Muslims: Identity, Culture, Politics and the Media*, Edinburgh: Edinburgh University Press 2010.

Stefano Kluzer is an economist (Bocconi University, Italy) with a Ph.D. in Information Systems from the London Schools of Economics and Political Science. He worked at the European Commission, JRC IPTS (Seville), Information Society Unit from 2007 to 2010, in charge of eInclusion research, specifically focusing on the use of ICT for/by immigrants and ethnic minorities. Between 2001 and 2006 he worked in Rome on various projects of the Italian e-Government Action Plan for regional and local authorities. From 1996 to 2001 he was with the regional development agency of Emilia-Romagna, ERVET Politiche per le Imprese Spa (Bologna) where he set up and coordinated the Information Society Department. Before then he worked as consultant and researcher on innovation policies and technology transfer to developing countries.

Pui-lam Law received his Ph.D. in sociology from the University of New South Wales, Australia, and is currently Assistant Professor in the Department of Applied Social Sciences of The Hong Kong Polytechnic University. His research interest is on modernity and social development in China. He co-authored *Marriage, gender, and sex in a contemporary Chinese village* (2004), co-edited *New technologies in global societies* (2006), and co-edited special issues on ICTs and China and on ICTs and migrant workers in contemporary China for Knowledge, Technology and Policy 21(1–2) (2008). Recently, he is working on ICTs and the issue of the identity of migrant workers in southern China.

Giuseppe Mantovani is full Professor of Psychology of Attitudes in the University of Padua, Italy, from 1990 to 2009, then appointed by the University of Padua to do research on interculture. He applied qualitative methods such as conversation analysis, discourse analysis, critical discourse analysis to issues of prejudice reduction and attitudes towards migrant people. Mantovani has published in journals such as Cultural Psychology; Cognitive Science; Human Relations; Mind, Culture and Activity. Books in English: Exploring borders (2000) and New communication environments (1996).

Daniel Miller is Professor of Material Culture Studies in the Department of Anthropology at University College London. His work on migrants includes the book Au pair written with Zuzana Burikova (2010) about the experience of Slovak au pairs working in London, and a forthcoming book written with Mirca Madianou called Migration and new media: Transnational families and Polymedia about the impact of new media on Filipina mothers' relationships with their left-behind children in the Philippines. Other books on the media include one on the use of the Internet in Trinidad (with Don Slater) and another on mobile phones in Jamaica (with Heather Horst). Other recent works include *Tales from Facebook* (2011), edited with Sophie Woodward, *Global denim* (Berg 2011), *Stuff* (2010), editor, *Anthropology and the individual* (2009) and *The comfort of things* (2008). His is currently carrying out research on migrants and their use of Webcams/Skype. He has also recently been involved in the establishment of a new program in Digital Anthropology at the Department of Anthropology at UCL.

Raul Pertierra (Editor) trained as an anthropologist and taught mainly in Australia (University of New South Wales). He specializes in Philippine society and culture and has written several books on migration, religion, politics, science and more recently on the use of mobiles and the Internet. He currently teaches at the University of the Philippines and at the Ateneo de Manila University. His latest research interest is in social networking and the uses of Web 2.0 among Filipinos.

Lilia Raycheva is Associate Professor at the Radio and Television Department of the Faculty of Journalism and Mass Communication at the St. Kliment Ohridski University of Sofia (since 1978). She has served as Vice Dean for Scientific Research and International Affairs and Head of Radio and Television Department, as a member (elected by the National Assembly) of the Council for Electronic Media—the regulatory authority for radio and television broadcasting in Bulgaria (2001–2008) and for three years as a member of the Standing Committee on Transfrontier Television at the Council of Europe (2005–2008). She lectures at home and abroad and has published extensively. Her professional portfolio includes a number of TV programs (one of them aired since 1980). She has also successfully participated in a number of international projects and networks on various mass media issues. Her scientific interests relate to information and communication technologies' impacts and media developments.

Alice Robbin is Associate Professor of Library and Information Science in the School of Library and Information Science at Indiana University Bloomington. She is co-director of the Rob Kling Center for Social Informatics. Her research interests include information policy, communication and information behavior in complex organizations and the societal implications of the information age. In addition to her research on the consequences of privacy law and policy for social research, the effects of digital inequality and information-seeking behavior on the Internet, she examines the political controversy over the federal reclassification of standards for racial and ethnic group data.

Polina Stoyanova completed her Bachelor's degree in Book Publishing and she has specialized in Media and Public Relations at the St. Kliment Ohridski University of Sofia. In 2008 she earned her Masters Degree of Electronic Media at the same University. She has been an MP assistant at the National Assembly and currently she is a Ph.D. student in the Book Publishing Department. Her research interests focus on book publishing, media and art.

Laura Suna, is research assistant at the University of Bremen, Germany. She works on the research project "Communicative connectivity of ethnic minorities: The integrative and segregative potential of digital media for diasporas" at the ZeMKI (Center for Media, Communication and Information Research). She studied Sociology at the University of Latvia in Riga and worked as a research assistant at the Advanced Social and Political Research Institute in Riga. Since 2007 she is a Ph.D. student at the University of Bremen. Topics of her Ph.D. project are media identities, popular media cultures and the potential of transcultural mediation between Latvian- and Russian-speaking youth in Latvia.

Manuel A. Vásquez teaches in the Department of Religion at the University of Florida. He has authored, co-authored and co-edited many books, including Globalizing the sacred: Religion across the Americas (2003), Immigrant faiths: Transforming religious life in America (2005), A place to be: Brazilian, Guatemalan, and Mexican immigrants in Florida's new destinations (2009), and most recently, More than belief: A materialist theory of religion (2011) and Living "illegal": The human face of unauthorized immigration (forthcoming).

Jane Vincent (Editor) joined the University of Surrey's Digital World Research Centre as a research fellow in 2002, prior to which she spent over twenty years in the U.K. mobile communications industry working for BT and O2 designing and implementing international mobile communications services in Europe. Jane's academic interests are in the user behaviors associated with mobile communications, in particular emotion and mobile phones, the topic of her research Doctorate and on which she has published numerous papers and book chapters, as well as jointly edited volumes including (with Leopoldina Fortunati) Electronic emotion: The mediation of emotion via information and communication technologies (2009).

Index